Fundamentals Of Packaging Dynamics

By
Richard K. Brandenburg, Ph.D
and
Julian June-Ling Lee, Ph.D.

School of Packaging
Michigan State University

Published in 2001 ©
by
L.A.B. Equipment, Inc.
1326 New Seneca Turnpike
Skaneateles, NY 13152 USA
and
1091 E. Green Street
Franklin Park, Illinois 60131 USA
www.labequipment.com

Printed in the United States of America

ISBN 0-9707313-0-2

ACKNOWLEDGEMENTS

Contributing editors:

Dr. James W. Goff
Professor - Michigan State University, East Lansing, Michigan

Dr. Marvin M. Kufahl
Professor - University of Wisconsin-Stout, Menomonie, Wisconsin

Mr. Robert M. Fiedler
Management Consultant, Minneapolis, Minnesota

Dr. John Kieskamp
Technical Director, Topa Verpakking, Voorhout, Netherlands

Creation and production of this textbook was sponsored by L.A.B., Skaneateles, New York. L.A.B. participation is the result of the company's interest in providing product designers and packaging engineers with information on the basics of packaging dynamics.

PREFACE

Let's face it ... no matter how brilliant the idea, nor ingenious the design ... no matter what the industry, nor how good the sales effort ... it's all a waste, if the customer receives a damaged product.

Consequently, package performance testing is no longer a goal that manufacturers dream about achieving some day, but an everyday tool used in nearly every industry in today's competitive world marketplace. Preshipment testing enables the most optimum product/package design combinations to be developed and verified, **before** losses occur.

A basic premise of good packaging design is that the product to be shipped is worthy of being packaged. Whenever possible, a product that is unrealistically fragile should go back to the design room for ruggedization. Costs are better put into the long-term product than the short-term packaging. Would you rather the manufacturer put the last 10% of its purchase price into your new product, or into its shipping container that has just been trucked to the landfill?

Once a manufacturer has a durable product, the distribution environment in which that product will be shipped must be carefully identified and detailed. Where and how will it be stored, loaded, transported, and unloaded? What will be the shipping configuration during each of the distribution steps? And what complex mixture of stresses, shocks, and vibrations will befall the product on the way to the customer?

Some packages/products are subjected to *compression* which is the loading force that occurs in stacked configurations and may or may not be evenly distributed. Others are *dropped* in the "pass, punt, and kick" contests on the loading docks of the world. Those that move on the rails are peculiarly likely to suffer shock damage upon *horizontal impact* from pallet marshalling or railcar switching. And every product everywhere is subjected to *vibration,* because the world doesn't react in a one for one manner.

According to the authors, "Fundamentals of Packaging Dynamics" was written to be an introduction to the dynamic behavior of products and packages in the distribution environment. It presupposes little or no background in shock and vibration. Simple mechanical and mathematical models are used throughout the book to lay a solid groundwork for later discussions of more complicated interactions between products, their packages, and the environments in which they both move.

I would encourage all engineers to learn these principles well. There are few jobs in which we are not called upon to use what can be gleaned herein. Even though we live in the days of government, industry, and consensus standardization, gone are the days of one test fits all. And here are the days of more and more stringent requirements that shippers warrantee both the safety of their cargo, and the safety of the people who may be exposed to hazardous contents should the packaging not be secure.

Good design engineering today takes a systematic approach that optimizes tradeoffs between product ruggedness and package protectiveness. Good manufacturing management demands that these tradeoffs be worked out in the company lab, **not** the world market.

TABLE OF CONTENTS

TABLES OF CONTENTS (CONTINUED)

FUNDAMENTALS OF PACKAGING DYNAMICS

1. INTRODUCTION: HAZARDS IN DISTRIBUTION

One of the goals of the packaging technologist is to get a product from one place to another without unacceptable loss or damage in an economic and practical manner. The product must be protected during distribution from damage or destruction and likewise, if the product is hazardous, people and property must be protected from the product. For consumer goods, the retailer expects to receive an attractive, unmarred, saleable package containing an undamaged product.

The package-product system should be designed to withstand the rigors of the distribution environment. From the point of the product's creation, the product-packaging system must survive the hazards encountered until it reaches its final destination, at which time the packaging may be discarded or recycled. The hazards of distribution are many and varied, and it is usually difficult or impossible to predict exactly what a product-package system is going to encounter.

Suppose, for example, that I wish to ship a device from my office at the University to a colleague in another state. The route that my package might take is shown in Figure 1.1. My package is first loaded onto a local truck and transported to the University's Stores-Receiving building where it is unloaded and stored. Shortly after, it is loaded onto another truck and carried to the local terminal which serves our metropolitan area. Again it will be off-loaded and moved about at the local terminal, until it is reloaded onto an "over the road" tractor-trailer truck. The package will be carried in the truck to the terminal serving my colleague's locality. Again it will be unloaded, moved about and reloaded onto one, or perhaps two, smaller trucks until it is delivered. Notice that during this seemingly simple distribution route the package is handled twelve to fifteen times and transported in four or five different trucks. In other, more complex distribution systems, a package may receive as many as 46 handlings as it is transported from plant to warehouse to consumer by rail, truck, "piggy back," ship and aircraft.

As a package moves through distribution, we need to be concerned with the dynamic forces encountered due to:

1. **Manhandling:** the dropping, throwing and other abuses applied by the manual loading, unloading and movement of packages.

2. **Warehouse handling equipment:** the stresses applied by mechanical handling equipment such as forklifts, conveyors, etc.

3. **Vehicle impacts:** the starting, stopping and other jolts due to the movement of trucks, railcars, ships and aircraft.

4. **Vehicle vibrations:** the naturally occurring vibrations resulting from the motion of engines and moving contact of the vehicle with highway and rails.

These four conditions result in impacts and vibrations to our product-package system. Package systems may be designed to minimize the damage caused in distribution, but sometimes the product must be redesigned in order to survive. In the next chapters we will examine these phenomena in greater detail.

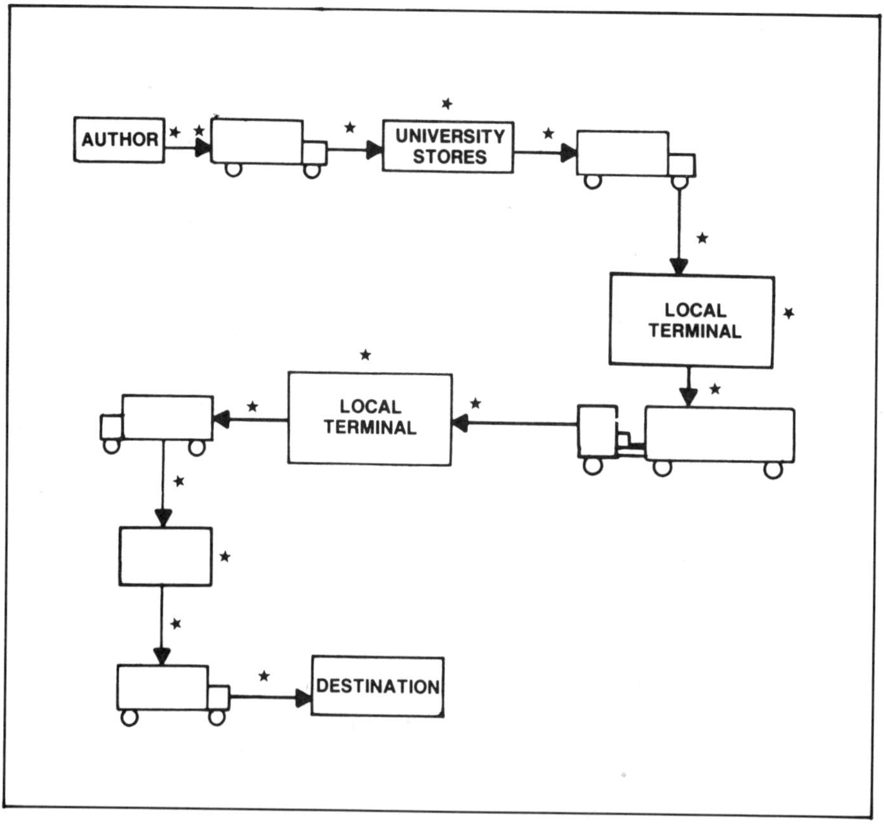

Figure 1.1. A Simple Distribution System (Each * Represents A Potentialy Hazardous Handling).

2. ACCELERATION

2.0 Fundamentals

The concept of acceleration is central to the understanding of the dynamic forces packages undergo in the distribution environment. To understand acceleration we must examine the simple principles[1] of space, time, and motion.

Let's examine the everyday act of driving an automobile. As we seat ourselves behind the wheel, the auto is at rest and our speed is zero. We turn the key, start the engine, release the emergency brake and engage the gears. As we push on the accelerator pedal we begin moving and our speed increases. When we reach our desired cruising speed, 35 miles per hour for instance, we slightly reduce the pressure on the accelerator to maintain constant speed. If we approach a stop sign, we remove our foot from the accelerator and depress the brake, reducing our speed until we come to a full stop.

There are a number of concepts in this example which are of interest:

> **Speed:** Speed is a measure of the distance covered by a moving object in a unit of time. Our automobile cruised at a speed of 35 miles per hour. So, if this cruising speed was maintained for two hours, we would have traveled 70 miles (2 hr x 35 miles/hr).
>
> **Velocity:** Velocity[2] is a measure of both the speed of an object and its direction. We may specify the velocity of our automobile by saying that it is traveling west at 35 miles per hour. The two vehicles shown in Figure 2.1 have identical speeds but different velocities.
>
> **Acceleration:** As we depressed the gas pedal we changed our velocity from zero to 35 miles per hour. The **rate** at which the velocity changes with respect to time is defined as the acceleration[3]. Suppose, in our example above, that it took one minute for our velocity to change from 0 to 35 mph. The average acceleration, then, may be computed from the definition:

$$\text{Acceleration} = \frac{\text{change in velocity}}{\text{change in time}} = \frac{35 \text{ mph-0 mph}}{\frac{1}{60}\text{hr}} = 2100 \text{ miles/hr}^2$$

> (Note that 1 minute was converted to 1/60 hour so as not to mix hours and minutes in our calculation.) This acceleration appears to be quite large, and indeed if we were able to maintain this level of acceleration for one hour we would be traveling at 2100 mph (or roughly three times the speed of sound). However, remember that this acceleration level was applied for only one minute, so that our final velocity of 35 mph was well within the speed limit.

---

[1]Newtonian physics will do for our study of package dynamics; packages rarely undergo relativistic motion, so it is unnecessary to involve the work of Dr. Einstein in this book.

[2]Velocity is a **vector** quantity. Vectors are used to indicate directions in three dimensional space.

[3]Acceleration is also a vector quantity.

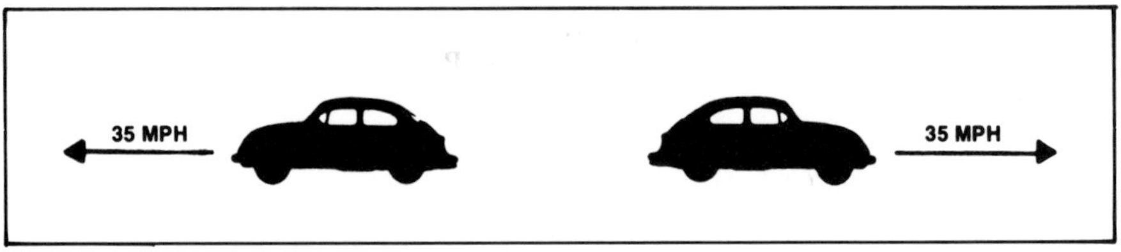

Figure 2.1. Identical Speed, Different Velocities.

The acceleration of an object is the rate at which its velocity changes with respect to time, and this change may be in the speed of the object, the direction of travel or both. The possibilities are illustrated in Figure 2.2.

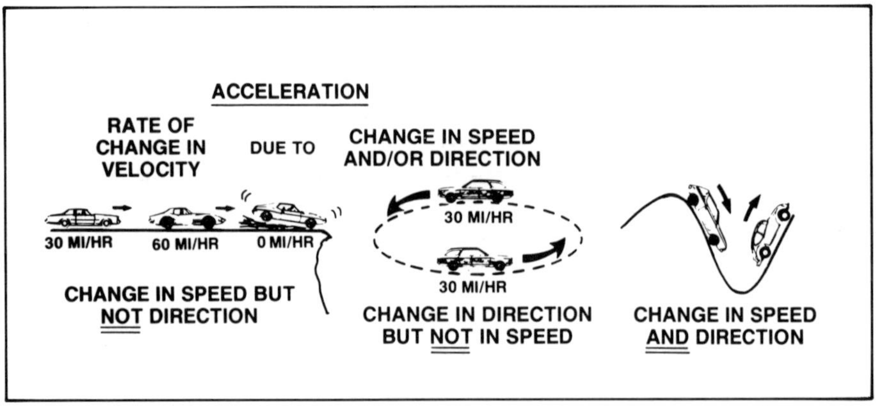

Figure 2.2. Accelerations

2.1 Mass

To complete our understanding of acceleration we must examine a fundamental property of matter. **Mass** is a measure of the amount of material composing an object and, more importantly, is the quality which causes an object to have **weight** when in a gravitational field. To understand this thoroughly we need to look at what happens when a body is dropped near the surface of the earth. Table 2.1 contains a history of the first one second of free fall for an object dropped near the earth's surface (ignoring the effects of air resistance).

TABLE 2.1
FREE FALLING BODY

Time Seconds	Milliseconds*	Distance Fallen		Velocity
0	0	0 in.	7.7 in. (0–0.2 sec)	0 in/sec. downward
0.1	100	1.9		38.6
0.2	200	7.7	23.2 in. (0.2–0.4 sec)	77.3
0.3	300	17.4		115.9
0.4	400	30.9		154.6
0.5	500	48.3		193.2
0.6	600	69.6		231.8
0.7	700	94.7		270.5
0.8	800	123.6		309.1
0.9	900	156.5		347.8
1.0	1000	193.2		386.4

*1 millisecond or 1 ms is equal to 0.001 second.

Notice that as the object falls, its instantaneous velocity increases[4], and it progressively covers greater distances. After the first 200 milliseconds the object has moved 7.7 inches downward, but during the next 200 milliseconds it moves 23.7 inches (30.9 in - 7.7 in). Our experience tells us that this continually increasing velocity is due to acceleration; an acceleration resulting from the attraction between the mass of the earth and the mass of the object. This attraction between the two masses is manifest as a force applied to the object; a force on the object which pulls (or pushes) it downward when the object is released.[5]

In fact this force due to the mass of earth is always present, so that whenever any object has the opportunity to fall towards the center of the earth, it will. This omnipresent force is commonly measured and called the object's weight.

Newton's second law of motion defines the force on a body as the product of the mass and the acceleration, or mathematically:

$$F = ma$$

(Force equals mass times acceleration)

Near the surface of the earth, where nearly all packaging activity takes place, the gravitational field is nearly uniform; that is, all objects experience the same pull of

---

[4]Instantaneous velocity is defined as the velocity at any one moment during an object's motion. If an object was moving at a constant velocity, the instantaneous velocity would be equal to the average velocity.

[5]The object also exerts an attractive force on the earth equal to the force exerted on the object by the earth. That is the earth and the object experience identical attractive forces, but because the mass of the object is infinitesimally small compared to that of the earth, the object moves toward the earth and the earth does not move toward the object. Incidentally, no one knows why masses are gravitationally attracted, but for packaging purposes we need not worry about the philosophical implications.

gravity. Consequently, we may measure the downward force exerted on an object by the earth by placing it on a spring scale. This force is the object's weight.

Weight = mass x gravitational acceleration

The gravitational acceleration, commonly referred to as "g", has been measured experimentally and found to be:

$g = 32.2\ \text{ft/sec}^2$ or

$g = 386.4\ \text{in/sec}^2$ or

$g = 9.8\ \text{m/sec}^2$ depending on which system of units you prefer.

"g" varies little from one spot on the earth's surface to another, so an object always has the same weight or experiences the same force wherever it is weighed.

It is obvious that weight is a particular result of gravitational attraction. If we were able to take an object to the surface of the moon, where, since the moon is considerably less massive than the earth, the gravitational acceleration is much smaller, we would find the object to weigh only one sixth of its earth weight. The physical properties of the object haven't changed, it still has the same mass it did on earth, but the attractive force exerted by the moon is much smaller than earth's. Likewise, an object in deep space with no other masses around it has a weight of zero, although its mass remains the same as it was on the earth or moon.

Back near the earth's surface all objects, including ourselves, experience a constant downward force. This constant force results in a constant weight and is due to the constant acceleration or pull of the mass of the earth. Everything here experiences a force of "one g"[6].

Suppose gravitational acceleration increased tenfold, so that everything experienced a constant force of 10 g's. This would mean that everything (and everyone) would "feel" as if it weighed ten times more than it did. Under 10 g's of acceleration a normally 200 lb. man would weigh 2,000 lb. The "10 g's" is, in fact, a way of expressing a multiplication of an object's weight due to an increase in acceleration. Increases in acceleration **do not** occur because the gravitational acceleration changes, but rather because of sudden changes in an object's motion. In packaging, these changes occur because of vibration and mechanical impact or shock, and often result in damage. Again, some familiar examples will help clarify these concepts.

Imagine we have a high powered sports car which will go from 0 to 60 mph in 5 seconds. What horizontal acceleration in g's will the driver experience? We'll assume for simplicity that the acceleration is constant[7] over the 5 seconds:

$$\text{Acceleration} = \frac{60\ \text{mph} - 0\ \text{mph}}{5\ \text{sec}} = 12\ \text{miles/hr} \cdot \text{second}$$

[6]Pronounced, "one gee".

[7]This is generally not true because for most engines there is a range of engine speeds for which power output, and therefore acceleration, is optimum.

Converting miles to feet and hours to seconds:

$$\text{Acceleration} = \frac{12 \text{ miles}}{\text{hr} \cdot \text{second}} \times \frac{5280 \text{ ft}}{\text{mile}} \times \frac{1}{3600 \text{ sec/hr}} = 17.6 \text{ ft/sec}^2$$

Comparing this to one g:

$$\frac{17.6 \text{ ft/sec}^2}{32.2 \text{ ft/sec}^2 \cdot \text{g}} = 0.55 \text{ g}$$

We see that the force experienced by the driver during the five seconds was approximately one half g. That is, if the driver weighed 200 lb., he would feel as if a weight of about 100 lb. was horizontally pressing him into the back of the seat.

As another example, consider the situation where the same automobile hits a telephone pole at 40 mph. During the crash, the auto goes from 40 to 0 mph in about 0.2 seconds. Computing the acceleration as before:

$$a = \frac{0 \text{ miles/hr} - 40 \text{ miles/hr}}{0.2 \text{ seconds}} = -294 \text{ ft/sec}^2$$

The negative sign indicates that we're dealing with deceleration. This deceleration or negative acceleration may be converted to g's:

$$a = \frac{294 \text{ ft/sec}^2}{32.2 \text{ ft/sec}^2 \cdot \text{g}} = 9.1 \text{ g's}$$

In this situation, our 200 lb. driver experiences for 0.2 seconds a force equal to 9.1 times his body weight (or 1820 lb.) trying to push him over the steering wheel and through the windshield, clearly a more dangerous situation than in the previous example.

2.2 Some Elementary Mathematics of Motion

Consider an object in space experiencing a constant acceleration, a, as shown graphically in Figure 2.3.

The average acceleration at any time, t, is equal to a. If, at some initial time t_i we start clocking the object (set $t_i = 0$) and note the initial velocity of the object (call it v_i), we would like to be able to predict its location and velocity at some later time (call that the final time or t_f). Figure 2.4 shows the relationship between velocity and time for our object.

We may derive an expression for the final velocity from the conditions and definition of acceleration:

$$a = \left(\frac{\text{change in velocity}}{\text{change in time}}\right) = \frac{v_f - v_i}{t_f - t_i}$$

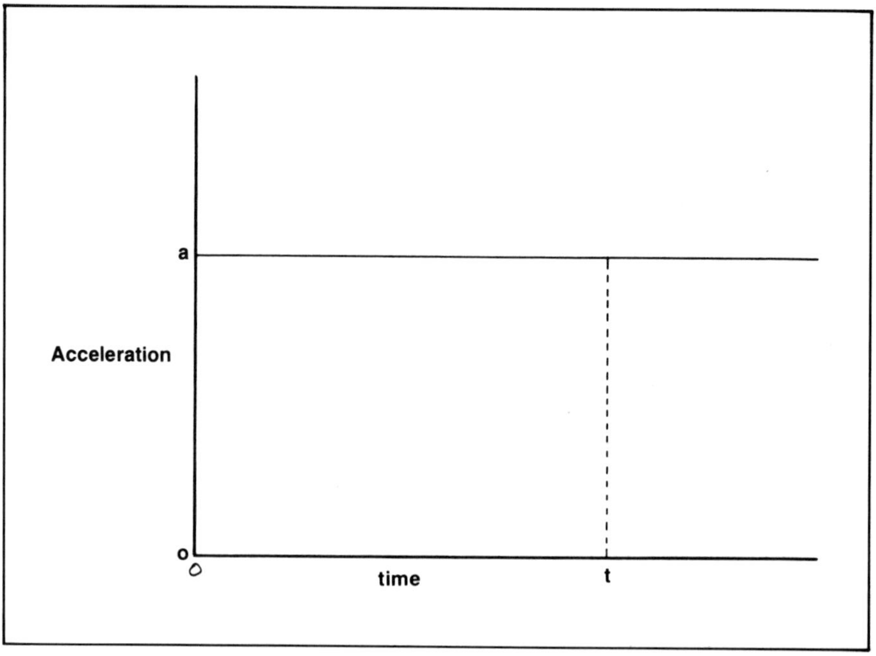

Figure 2.3. Constant Acceleration.

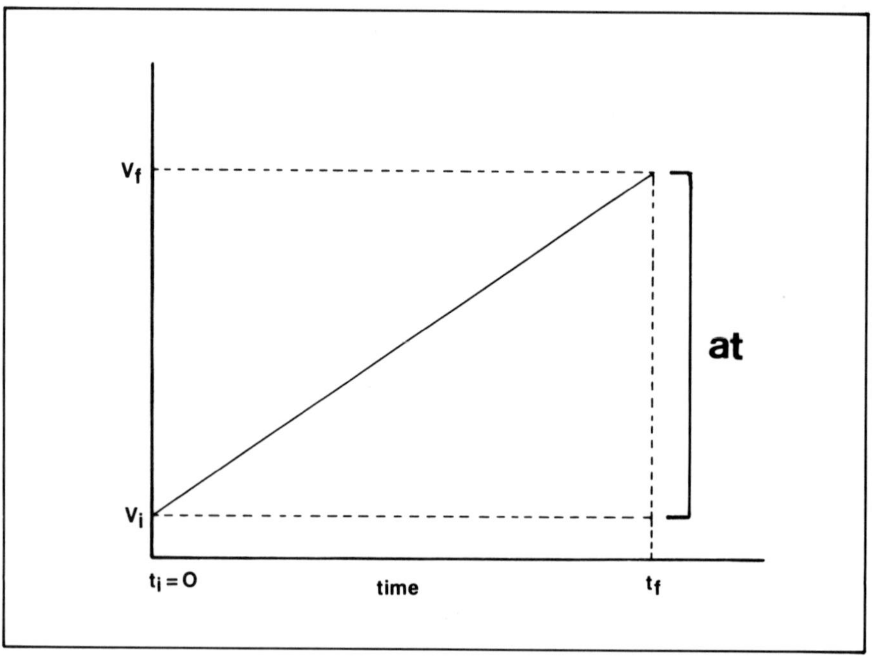

Figure 2.4. Velocity Due to Constant Acceleration.

and, since $t_i = 0$, $a = \frac{v_f - v_i}{t_f}$

so $at_f = v_f - v_i$

then, $v_f = v_i + at_f$ 2.1

If a is constant, velocity changes uniformly with time. The average velocity is given by:

$$\bar{v} = \frac{(v_i + v_f)}{2} \qquad 2.2$$

If the object's position at $t_i = 0$ was noted to be at coordinate x_i, then its position later at time t_f may be found,

$t_i \rightarrow \ldots\ldots\ldots t_f$

$x_i \qquad\qquad\qquad x_f$

The final coordinate x_f is equal to the sum of the initial coordinate plus the product of the average velocity and the time;

$$x_f = x_i + \bar{v}\,(t_f - t_o) = x_i + \bar{v}t_f = x_i + 1/2\,(v_i + v_f)\,t_f \qquad 2.3$$

Substitute Equation 2.1 into 2.3 and we have a useful result;

$$x_f = x_i + 1/2\,(v_i + v_i + at_f)\,t_f$$

$$= x_i + 1/2\,(2v_i + at_f)\,t_f$$

$$= x_i + v_it_f + 1/2\,at_f^2$$

We may simplify this by noting that the time elapsed is

$t = t_f - t_i = t_f$

then

$$x_f = x_i + v_it + 1/2\,at^2 \qquad 2.4$$

Equations 2.1 and 2.4 provide us with the capability[8] of predicting position and velocity from elapsed time for an object under constant acceleration. Because the

[8]One additional relationship may be found by solving 2.1 for t and substituting the result into 2.3:

$v_f = v_i + at$, $t = \frac{v_f - v_i}{a}$

$x_f = x_i + 1/2\,(v_i + v_f)t = x_i + 1/2\,(v_i + v_f)\left(\frac{v_f - v_i}{a}\right)$

$= x_i + \frac{1}{2a}(v_f^2 - v_i^2)$; $v_f^2 - v_i^2 = 2a\,(x_f - x_i)$

Then: $v_f^2 = v_i^2 + 2a\,(x_f - x_i)$ 2.5

earth's gravity provides constant acceleration we may alter our equations to apply to objects dropped above the earth's surface by simply substituting g for a.

$$a = g$$

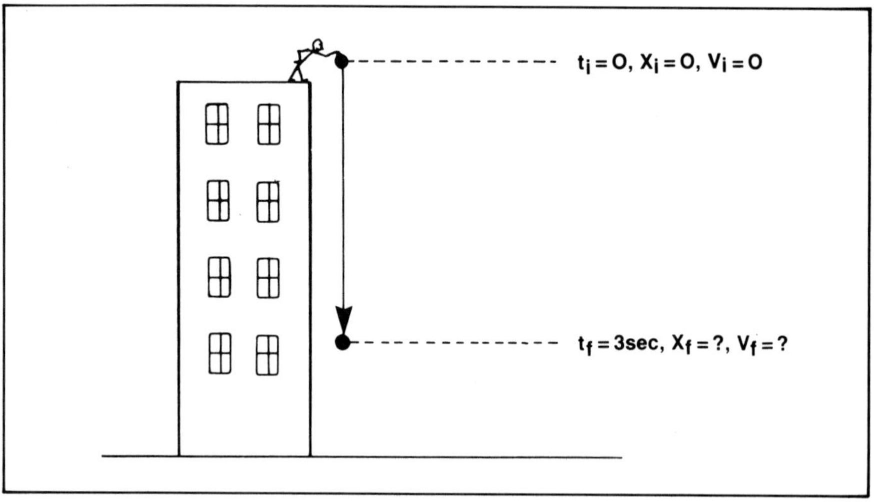

Figure 2.5. Dropping an Object Off a Building.

If we drop a marble from a high building, what will its position and velocity be three seconds later?

We set the top of the building as our zero reference point and start our clock at $t_i = 0$ when we release the marble. As we release the marble its initial velocity, v_i, is zero.

What is the position relative to the top of the building three seconds later? Using Equation 2.4,

$$x_f = x_i + v_i t + 1/2\ at^2$$

becomes

$$x_f = 0 + (0 \cdot 3\text{ sec}) + (1/2)\ (32.2\text{ ft/sec}^2)\ (3\text{ seconds})^2$$

$$x_f = 144.9\text{ ft.},$$

that is, the marble has fallen 144.9 feet measured from the top of the building. Its velocity at that point and moment may be found using Equation 2.1:

$$v_f = v_i + at = 0 + gt = (32.2\text{ ft/sec}^2)(3\text{ sec}) = 96.6\text{ ft/sec (downward)}$$

2.3 Packaging Application

We may now use equations developed in the previous section to characterize what happens when a package is dropped and hits the floor. Figure 2.6 illustrates the typical situation where the drop height, h, is known.

The velocity at which the package impacts the floor is related to package and product damage. We would like to be able to estimate that **impact velocity** from an easily measured parameter, such as the drop height. Such an expression **may be** developed from Equation 2.4.

$$x_f = x_i + v_i t + 1/2\ gt^2$$

where the initial velocity is zero.

$$x_f = x_i + 1/2\ gt^2 \qquad \textbf{2.5}$$

Note that the drop height is equal to the difference between the final and initial coordinates.

$$h = x_f - x_i = 1/2\ gt^2$$

so

$$h = 1/2\ gt^2$$

and

$$t = \sqrt{\frac{2h}{g}} \qquad 2.6$$

where t is the time it takes the package to fall a distance, h.

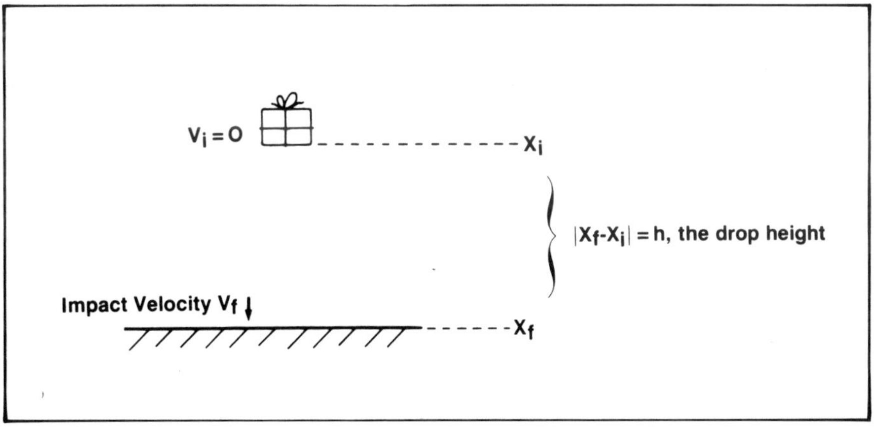

Figure 2.6. A Known Drop Height.

Now that we know the drop time, it is a simple matter to use Equation 2.1 and find an equation for the impact velocity.

Impact velocity = final velocity,

$$v_f = v_i + gt$$

Again v_i, the initial velocity, is zero,

$$v_f = gt$$

substituting Equation 2.6 for t,

$$v_f = g\sqrt{\frac{2h}{g}} = \sqrt{2gh}$$

Let's indicate the impact velocity as v_I, so,

$$v_I = \sqrt{2gh} \qquad 2.7$$

Suppose a package falls from the top of a four foot stack; what is the impact velocity?

$$h = 4 \text{ ft}$$

$$v_I = \sqrt{2gh}$$

$$v_I = \sqrt{2 \cdot 32.2 \text{ ft} \cdot \text{sec}^2 \cdot 4 \text{ ft}}$$

$$v_I = 16 \text{ ft/sec (11 mph)}$$

How long did it take the package to fall this distance?

$$t = \sqrt{\frac{2h}{g}}$$

$$t = \sqrt{\frac{2 \cdot 4 \text{ ft}}{32.2 \text{ ft/sec}^2}}$$

$$t = 0.50 \text{ seconds (500 milliseconds)}$$

If the package falls on a concrete floor it has about 0.003 seconds (3 milliseconds) to change its velocity from 16 ft/sec to zero. What acceleration (deceleration) is involved?

Recalling that acceleration is the change in velocity divided by the change in time,

$$\text{Acceleration} = \frac{16 \text{ ft/sec - 0}}{0.003 \text{ sec}} = 5333 \text{ ft/sec}^2$$

$$\text{or} \quad \frac{5333 \text{ ft/sec}^2}{32.2 \text{ ft/sec}^2 \text{ g}} = 166 \text{ g's}$$

In this example we have a simple, common four foot drop. Our package, due to the acceleration of gravity, reaches a velocity of 16 ft/sec (or 192 in/sec) a moment before the impact with the concrete floor. To slow down and come to a dead stop in 0.003 seconds, the package undergoes an acceleration of 166 g's; that is, for part of the deceleration time, the package "feels" as though it weighs 166 times what it does under ordinary circumstances. Not many articles, manmade or otherwise, can withstand this level of acceleration even for short periods of time.

PROBLEM SET, CHAPTER 2

1. An object is dropped in the vicinity of the earth's surface. Find its position relative to the release point and its velocity after: 1 second, 5 seconds, and 10 seconds. ($g = 32.2$ ft/sec^2)

2. An object is dropped. At what distance from the release point does it reach the following speeds: 24 ft/sec, 55 ft/sec, 100 ft/sec, 200 ft/sec?

3. Find the time required for an object to fall a distance of 6", 18", 36", and 60", .75 meters, 2.25 meters. (For the last two, $g = 9.8$ m/sec^2, 1 meter = 39.37 in.)

4. Find the velocity of an object at the instant of contact with a surface (floor or ground) when it is dropped from each of the heights listed in Problem #3.

5. A package falls from a four and one half foot high stack. How many seconds elapse until it hits the floor? At what velocity (expressed in inches/second) is the package traveling at the instant it hits the floor?

3. VIBRATION

3.0 Simple Vibratory Motion

Vibration is something we encounter in everyday life. As we wait by the railroad crossing, we can feel the ground vibrating from the train's motion. Riding in an automobile we feel the vibrations caused by the road's surface. In an airplane we feel the engines vibrate. Noise is transmitted by vibration of the air, and is turned into a signal recognized by our brain through the vibration of the eardrum. The heat of any object is merely the vibratory motion of its molecules and atoms. Even visible light is a form of vibration[1]. Not surprisingly, packaged items passing through the distribution system encounter vibrations. It is these complex mechanical vibrations and the accompanying changes in acceleration which may lead to physical damage.

Before we can deal with the complex mechanical vibrations found in the distribution environment, we need to examine the simplest form of vibration, **simple harmonic motion.** Figure 3.1 shows a simple harmonic oscillator; a mass attached to a frictionless spring, vibrating in the absence of a gravitational field.

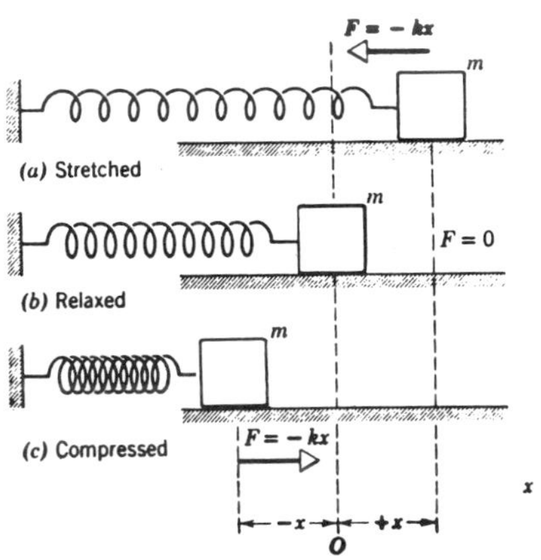

Figure 3.1. A Simple Harmonic Oscillator (5).

Note that in "a" the spring is stretched and the force of the spring is pulling the mass to the left. In "c" the spring is compressed and pushing the mass to the right. At "b" the mass is at the midpoint of the motion and is being neither pushed nor pulled. Imagine the motion as "a-b-c" repeated over and over again. This is harmonic motion.

---

[1]Electromagnetic vibration.

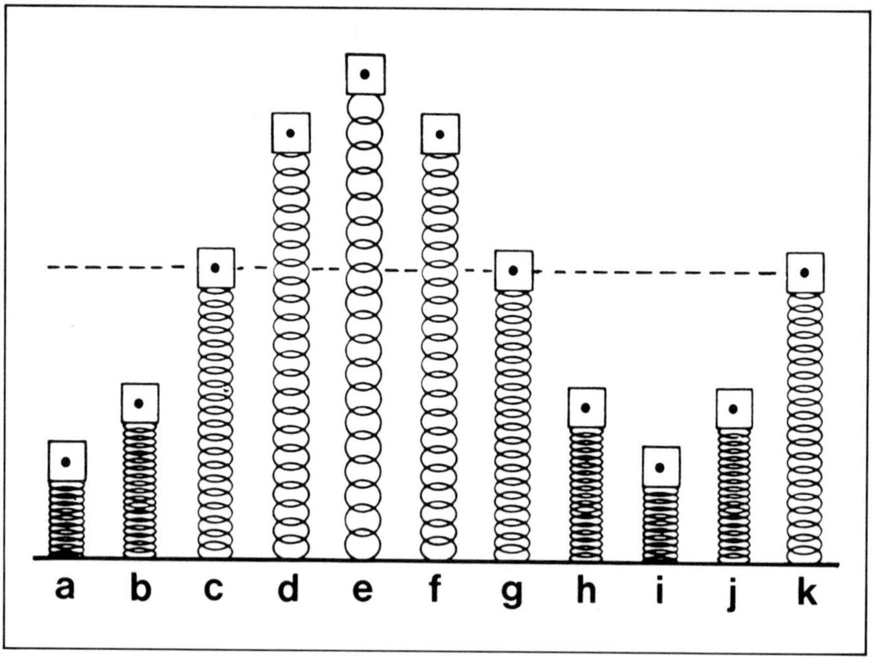

Figure 3.2. A Vertical Harmonic Oscillator.

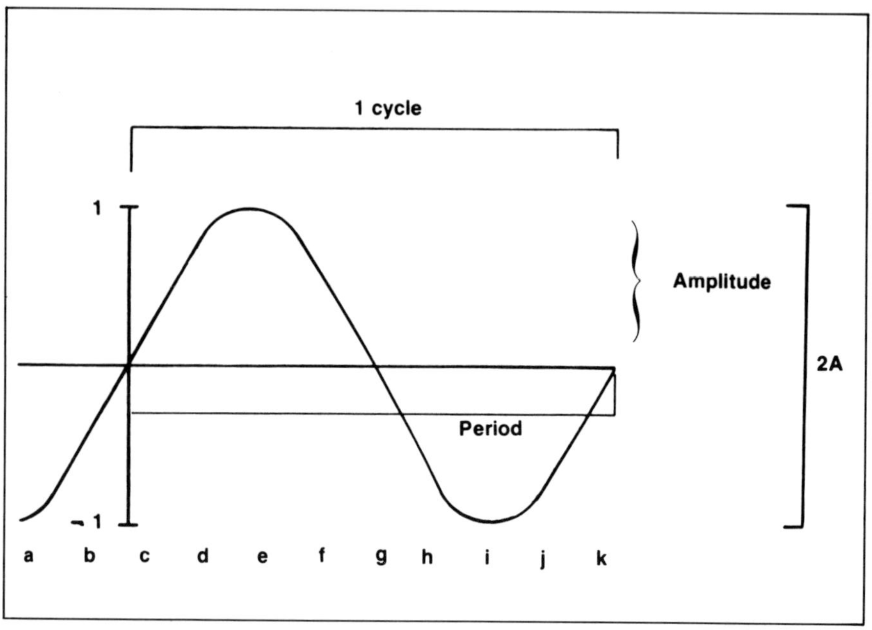

Figure 3.3. A Graph of Motion of the Harmonic Oscillator as a Function of Time.

Let's move this simple harmonic oscillator into our frame of reference by rotating it 90° and attaching it to the floor. Now depress the spring-mass, release it, and observe the motion as is shown in Figure 3.2. Some arbitrary coordinates are drawn on the vertical axis to help measure the travel of the mass during its vibration.

In Figure 3.2a the mass has been depressed to one inch below the rest or **equilibrium** position and released. The mass moves upward through positions b, c and d until it reaches a maximum extension at point e. It moves downward through positions f, g and h until at i it is again at its lowest point, and the process begins again.

If we graph the position of the black dot on the mass in Figure 3.2 we get a plot of the motion versus time, as shown in Figure 3.3. Several important parameters are illustrated:

PERIOD: The period is the time required for one complete cycle. If, for our example, the time elapsed between points c and k is measured to be one half second, the period of the vibration is said to be one half second. For shorthand purposes:

T = period in seconds

FREQUENCY: The frequency is defined as the number of cycles occurring within a specified unit of time. The frequency and the period are reciprocally related:

$$\text{Frequency (cycles/second)} = \frac{1}{\text{period (seconds)}} \qquad 3.1$$

A frequency of one cycle per second is defined as one Hertz or 1 Hz. The example above has a frequency of 2 Hz. We will also denote frequency as "f". Equation 3.1 may then be expressed as:

$$f(hz) = \frac{1}{T(sec)} \qquad 3.2$$

AMPLITUDE: The amplitude is the displacement of the mass relative to the zero or equilibrium point. For our example, the amplitude was one inch. This amplitude, denoted by "A", is often referred to in shock and vibration literature as the "single" amplitude or "zero to peak" amplitude. Measurement from the lowest position of a vibrating mass to its topmost position, 2 inches in our example, is defined as the "double amplitude" or "peak to peak" amplitude[2].

It is convenient to mathematically represent the motion of the vibrating spring-mass system with the sine function:

$$x = A \sin (pt) \qquad 3.3$$

---

[2]Sometimes it is easier to measure the double amplitude of a vibrating system and divide by 2 to get A.

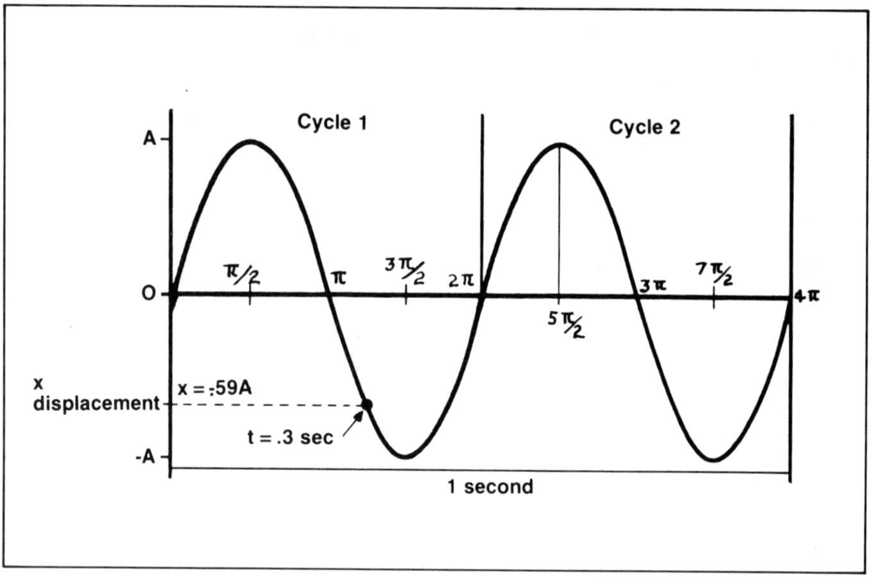

Figure 3.4. A Sine Function: The Displacement of an Oscillating Mass from the Equilibrium Position. (See Equation 3.3)

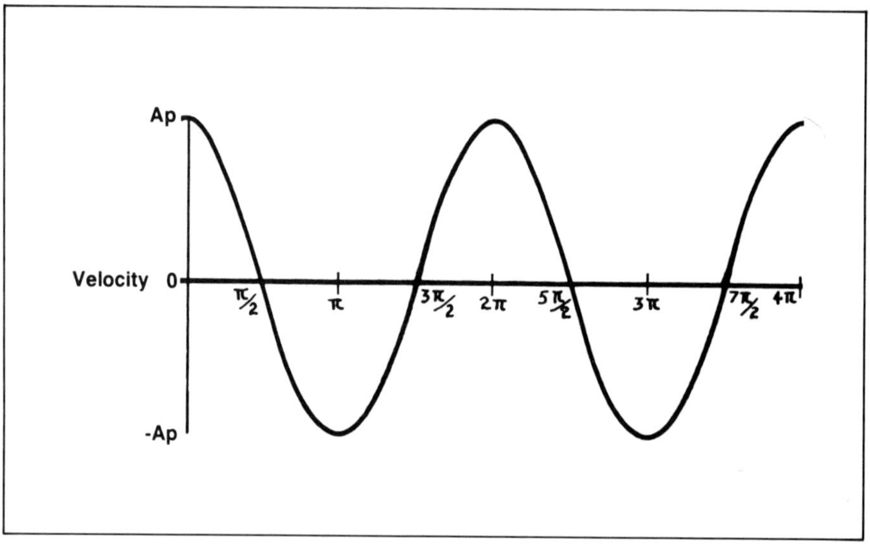

Figure 3.5. The Velocity of a Vibrating Mass. (See Equation 3.5)

where x is the displacement of the mass measured from the equilibrium position at time t_i, as shown in Figure 3.4. In the example shown, the vibration frequency is 2 Hz (two cycles per second) and the horizontal axis is scaled in radians.[3] The circular frequency, p, is defined:

$$p = 2\pi f \qquad 3.4$$

where f is the frequency of vibration. For our example $P = (2\pi \cdot 2 \text{ Hz}) = 4\pi$ cycles/second, and at time t = 0.3 seconds, for instance, the displacement may be calculated:

$$x = A \sin (4\pi \text{ cycles/sec} \cdot 0.3 \text{ sec})$$

$$= -0.59 \text{ A (about 59\% of its maximum, one inch downward travel)}$$

The ability to calculate the position of the vibrating mass is of little value. We know that it is somewhere between +A and -A, which is adequate for packaging purposes. What is important is the velocity and acceleration of the mass during this vibration. The velocity of the mass is defined as the change in displacement divided by the change in time or

$$v = \frac{\text{change in x}}{\text{change in time}} = \frac{dx}{dt}^{4}$$

$$v = \frac{d}{dt}(A \sin pt) = A\, p \cos (pt) \qquad 3.5$$

Figure 3.5 is a graph of the velocity of the vibrating mass using the same horizontal scale as in Figure 3.4.

Carrying the same procedure one step further, we may find the acceleration, a:

$$a = \frac{\text{change in velocity}}{\text{change in time}} = \frac{dv}{dt}$$

$$a = \frac{d}{dt}(A\, p \cos (pt))$$

$$a = -A\, p^2 \sin (pt) \qquad 3.6$$

Figure 3.6 illustrates the acceleration of the vibrating mass using the same horizontal scale as in Figures 3.4 and 3.5.

Notice from Figures 3.4, 3.5 and 3.6 that as the displacement x changes, the velocity and acceleration change. These changes in motion are summarized in Figure 3.7

---

[3] Recall that $\pi/2$ radians = 90°, π radians = 180°, etc.

[4] Because x is a complex function we must use a calculus concept, differentiation, to develop expressions for velocity and acceleration.

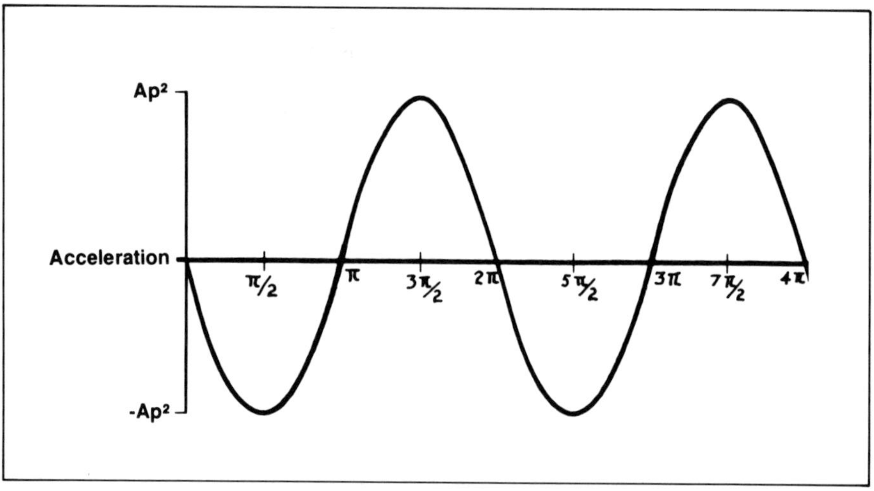

Figure 3.6. The Acceleration of a Vibrating Mass. (See Equation 3.6)

where one cycle of the vibrating mass is shown. At point A the mass is passing through the equilibrium point and its velocity has reached a maximum. From Equation 3.5 the velocity may be calculated, setting t = 0:

$$v = A\ p \cos(pt) = A\ p \cos(0) = A\ p$$

The maximum value of the velocity is equal to the product of the amplitude and the circular frequency. At this same point the acceleration is at a minimum. From Equation 3.6:

$$a = -A\ p^2 \sin(pt) = -A\ p^2 \sin(0) = 0$$

At point B the spring is stretched and is exerting a force, and therefore an acceleration in the downward direction. This slows the mass down; its velocity is decreasing. At point C the downward acceleration produced by the extension of the spring has reduced the velocity of the mass to zero:

$$v = A\ p \cos(pt) = A\ p \cos(4\pi \cdot 0.125\ \text{sec}) = 0$$

The acceleration, however, is at a maximum, pointing in the downward direction (hence the negative sign).

$$a = -A\ p^2 \sin(4\pi \cdot 0.125\ \text{sec}) = -A\ p^2$$

This negative or downward acceleration reverses the motion of the mass, so that at point D it is on its way down toward the equilibrium point again. At point E, at its passage through the equilibrium point, the acceleration has decreased to its minimum value, zero, and the velocity is at a maximum,

$$v = -A\ p,$$

the negative sign indicating its downward direction. As the mass at point F approaches the bottom of its travel, the spring is compressing and is resisting the downward motion, applying an acceleration in the upward or positive direction, until at point G the mass once again stops (velocity equals zero) and reverses its direction under the maximum (positive) value of the acceleration. The motion then continues in the upward direction until once again the mass is at the equilibrium position and the cycle begins again.

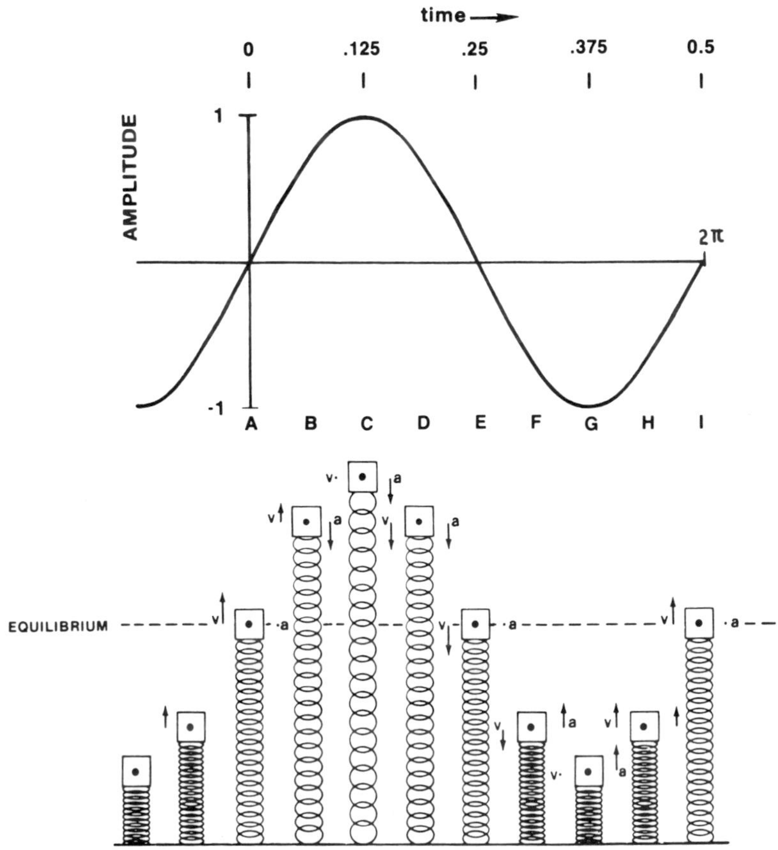

	A	B	C	D	E	F	G	H	I
VELOCITY, v	MAX		MIN		MAX		MIN		MAX
ACCELERATION, a	MIN		MAX		MIN		MAX		MIN

Figure 3.7. Free Vibration at 2 Hz.

This analysis has uncovered three important points:

1. The velocity of the mass reaches a maximum value as the mass passes through the midpoint or equilibrium point of its motion ($x=0$), and has a value of zero at the extremes of the motion ($x=\pm A$). (Velocity is said to be 90 degrees out of phase with displacement.)

2. Acceleration reaches a maximum value at the extremes or endpoints of motion where the displacement is the greatest and velocity is zero, and reaches its minimum value at the midpoint of the motion ($x=0$) where velocity is at a maximum. Acceleration has its greatest negative value ($-A P^2$) at the point where displacement has the greatest positive value ($x = A$): Acceleration and displacement are said to be 180° out of phase.

3. The maximum values for displacement, velocity and acceleration are, respectively:[5]

$$x_{max} = A \quad 3.7$$

$$v_{max} = A\,p \quad 3.8$$

$$a_{max} = A\,p^2 \quad 3.9$$

It is the large accelerations experienced at the extremes of the motion which are related to vibration damage in package distribution.

3.1 The Yo-Yo Analogy

A child's yo-yo undergoes a motion which is very similar to the simple harmonic motion described in the previous section. The yo-yo's motion is shown in Figure 3.8.

Let's start with the yo-yo at its highest position. As time goes by the yo-yo drops down, passing an arbitrary reference point (position B) to its lowest point (C). From there the yo-yo moves upward, passing the reference point again (D), ending up at the highest point again before it moves downward to repeat its periodic motion. The time required to complete one periodic motion (one cycle) is termed the period (T), usually measured in seconds. The number of cycles completed in one second is termed the frequency (f) in units of Hertz (Hz). Frequency is related to period by:

$$f = \frac{1}{T} \quad \text{(T in seconds)}$$

---

[5]These maxima are found from Equations 3.3, 3.5 and 3.6 by taking the absolute value of each function. Recall that the largest value of the cos or sin is 1:

$$x_{max} = |x| = |A \sin(pt)| = |A| = A, \; |\sin(pt)|_{max} = 1$$

$$v_{max} = |v| = |A\,p \cos(pt)| = |Ap| = A\,p, \; |\cos(pt)|_{max} = 1$$

$$a_{max} = |a| = |-A\,p^2 \sin(pt)| = |-A\,p^2| = Ap^2$$

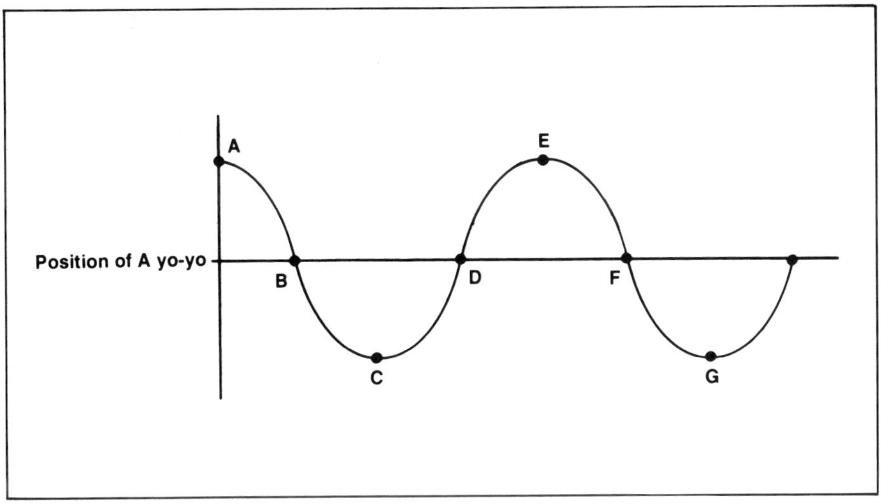

Figure 3.8. The Motion of A Yo-Yo.

The frequency at which the yo-yo vibrates, as played by a normal child, is probably in the neighborhood of 1 Hz. If observed closely, the speed at which the yo-yo moved is changing all the time.

At position A the speed is zero.

From A to B the speed is increasing; at B the speed is at a maximum.

From B to C the speed is decreasing so it can stop at C (where speed is zero again).

From C to D the speed increases.

From D to E the speed reduces again to a full stop at E.

Of more interest is the constant change of acceleration during the ups and downs. Let's start with Point C. A jerk is applied at the instant the yo-yo is at position C. Jerk is a force which causes the yo-yo to accelerate upward. The intensity of the jerk decreases from C to D. Therefore, the acceleration gets smaller and smaller and at D there is no acceleration (ignoring the omnipresent one g due to gravity). From D to E the yo-yo is decelerating at an increasing rate, with the maximum deceleration at E. The acceleration behavior of the yo-yo from A to C is the same as C to E except the direction of acceleration.

The important thing is that acceleration in vibration changes constantly and it has maximum and minimum values. The motion of a yo-yo is a close approximation of the vibration of a mass attached to a linear spring; the simple spring-mass system discussed earlier.

3.2 The Linear Spring

We ultimately need to examine what happens to an object, packaged or unpackaged, in the distribution environment when it is subjected to outside vibrations. The mass on a spring, introduced in Section 3.0, is the ideal model for a product on a cushion, or a component within a product or any number of other situations where we would like to predict behavior resulting from distribution vibrations. To simplify things we will assume the spring is a massless[6] linear spring, and our model will look like that shown in Figure 3.9.

The linear spring is used because it has very simple properties. The load displacement curve for a linear spring is shown in Figure 3.10.

A **LINEAR SPRING** is a spring that deflects (either in compression or elongation) proportionally to the force applied to it. Suppose we place a weight W_1 on the linear spring shown in the left half of Figure 3.9. The spring will compress a distance of x_1, measured from the unloaded free height. Now remove W_1 and replace it with a second weight W_2 which weighs, for example, three times W_1 ($W_2 = 3W_1$). Our linear spring will compress a distance x_2 which is three times x_1 ($x_2 = 3x_1$). The load-displacement curve for such a spring is then a straight line (Figure 3.10), hence the term linear. The slope of the load-displacement curve is k, the **spring constant,** where k may be computed from the expression:

$$k = \frac{W_2 - W_1}{x_2 - x_1}, \qquad 3.10$$

and is usually given in pounds per inch (lb/in). The force exerted by the spring upward to counteract the downward pull of gravity on the mass is given by:

$F_s = kx$, where x is the displacement.

For a specific mass on a specific linear spring, we may measure the downward displacement when the system is at rest. This is the **static deflection,** δ_{st} (pronounced, delta-s-t):

$$\delta_{st} = \frac{W}{k} \qquad 3.11$$

The static deflection of a product on a cushion may be simply measured in the laboratory and will give us useful information on the dynamic behavior of the product cushion system when its subjected to external vibrations.

3.3 The Natural Frequency

When we displace the mass of the spring-mass system downward and release it, it will oscillate according to the displacement equation:

---

[6]The weight of the spring is very small compared to the weight of the object or product.

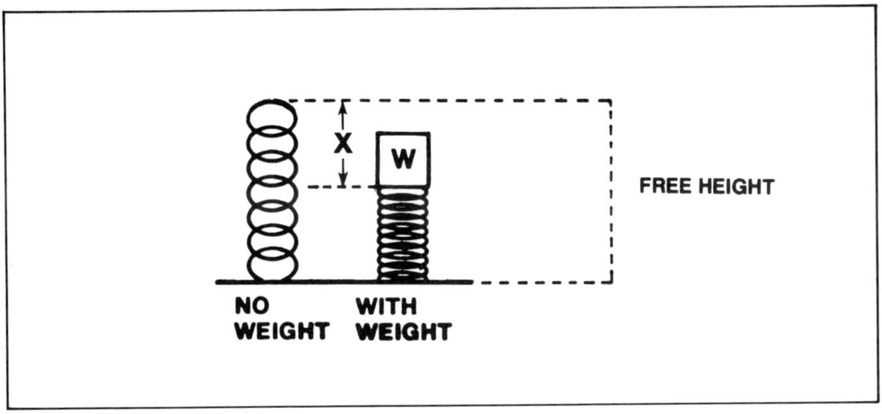

Figure 3.9. A Mass on a Linear Spring.

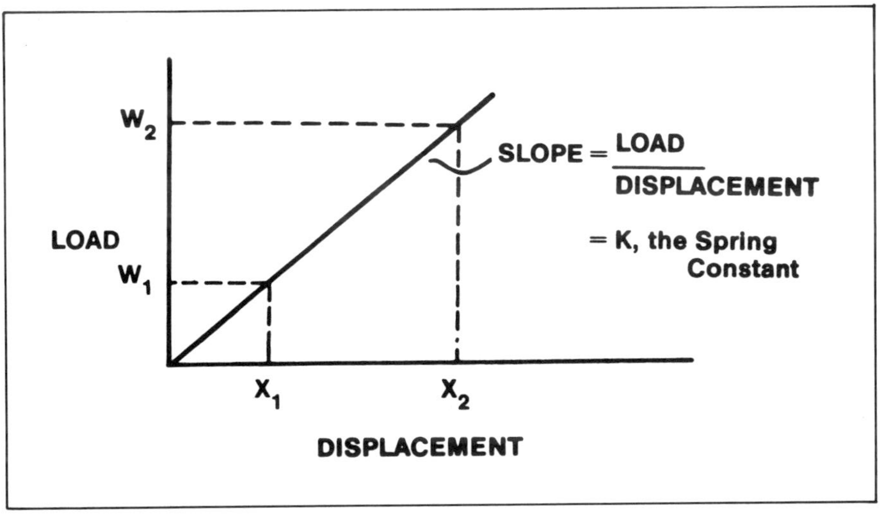

Figure 3.10. The Load-Displacement Curve for a Linear Spring.

$$x = A \sin (pt)$$

and the velocity and acceleration equations developed in Section 3.0. In this situation, there is a definite relationship between the circular frequency, p, the linear spring constant, k, and the weight of the mass, W:

$$p^2 = \frac{kg}{W} \qquad 3.12$$

where g is the acceleration of gravity (386.4 in/sec^2).

Recall that the circular frequency is related to the frequency of the vibrating system:

$$p = 2\pi f \qquad 3.13$$

Our spring-mass system will always vibrate with the same frequency once it is set into free motion. That is, when a force is applied to the system setting it in motion and then withdrawn, the system will vibrate at a precise frequency. This frequency is defined as the **natural frequency,** denoted by f_n. Equation 3.13 then becomes:

$$p = 2\pi f_n \qquad 3.14$$

Eliminating p from Equations 3.12 and 3.14 we may find an expression for the natural frequency:

$$p^2 = \frac{kg}{W}, \quad p = \sqrt{\frac{kg}{W}}$$

Since $p = 2\pi f_n$, then

$$2\pi f_n = \sqrt{\frac{kg}{W}} \qquad \text{and,}$$

$$f_n = \frac{1}{2\pi}\sqrt{\frac{kg}{W}} \qquad 3.15$$

The natural frequency is now known for a linear spring mass-system if we have values for the spring constant and the weight. For example, if:

W = 10 lbs. and k = 500 lb/in, then,

$$f_n = \frac{1}{2\pi}\sqrt{\frac{(500 \text{ lb/in})\,(386.4 \text{ in/sec}^2)}{10 \text{ lb.}}}$$

$$= 22 \text{ cycles/sec or } 22 \text{ Hz}$$

Equation 3.15 may be simplified by combining the constants 2π and g:

$$f_n = \frac{1}{2\pi}\sqrt{\frac{kg}{W}} = \frac{\sqrt{g}}{2\pi}\sqrt{\frac{k}{W}} = \frac{\sqrt{386.4}}{2\pi}\sqrt{\frac{k}{W}}$$

$$= 3.13\sqrt{\frac{k}{W}} \qquad 3.16$$

so, $f_n = 3.13\sqrt{\frac{k}{W}}$, where k is in units of lb/in and W is expressed in pounds. This relationship may be simplified even further by recalling the expression for the static deflection, Equation 3.11:

$$\delta_{st} = \frac{W}{k}$$

Take the reciprocal of δ_{st} and substitute in Equation 3.16:

$$\frac{1}{\delta_{st}} = \frac{k}{W}, \text{ and}$$

$$f_n = 3.13\sqrt{\frac{k}{W}} = 3.13\sqrt{\frac{1}{\delta_{st}}} \qquad 3.17$$

where δ_{st} is expressed in inches. Now we have a relationship between the static deflection of the spring-mass system and its natural frequency. Table 3.1 contains some static deflection values and the corresponding natural frequencies. As the deflection decreases, the natural frequency of the system increases, and conversely as the natural frequency of the system decreases, the deflection increases. This relation holds not only for simple spring-mass systems but, more importantly, also for products on cushions or structural components in products. As we shall see, this has some serious consequences for packages in the distribution system.

TABLE 3.1

δ_{st} Inches	f_n Hertz
2	2.2
1	3.2
0.5	4.4
0.25	6.3
0.10	9.9
0.01	31
0.001	99

3.4 More on the Vibrating Spring-Mass System

At this point we may eliminate the circular frequency parameter, p, and examine the velocity and acceleration relationships. Recall from Equations 3.12 and 3.13 that p may be expressed as:

$$p = \sqrt{\frac{kg}{W}} \text{ or } p = 2\pi f$$

where $f = f_n$ for the freely vibrating spring-mass system. We also know from our earlier examination of sinusoidal motion that the maximum velocity reached by the vibrating mass and the maximum acceleration experienced by the mass are given as:

$$v_{max} = Ap \qquad 3.8$$

$$a_{max} = Ap^2 \qquad 3.9$$

where A is the single amplitude of the motion of the mass. For packaging we are concerned with the maximum velocity and, more importantly, with the maximum acceleration of a product. A set of equivalent expressions may be developed for the maximum velocity:

$$v_{max} = A\sqrt{\frac{kg}{W}} \qquad 3.18$$

or

$$v_{max} = A\,(2\pi f_n) \qquad 3.19$$

and for the maximum acceleration:

$$a_{max} = A\,\left(\frac{kg}{W}\right) \qquad 3.20$$

or

$$a_{max} = A\,(2\pi\,f_n)^2 \qquad 3.21$$

To see how all of these parameters and functions are interrelated, let's take an example. Suppose we have a 10 lb. product supported by a piece of cushion material, a cushion which we will assume to behave as a linear spring. At rest the cushion is depressed 0.61 inches from the unloaded condition. At what frequency will this system vibrate if it is set into motion?

We can find the natural frequency from Equation 3.17:

$$f_n = 3.13\sqrt{\frac{1}{\delta_{st}}} = 3.13\sqrt{\frac{1}{0.61}} = 4\text{ Hz}$$

Because we know the product's weight we may also find the linear spring constant from Equation 3.11:

$$\delta_{st} = \frac{W}{k}\ ;\ k = \frac{W}{\delta_{st}} = \frac{10\text{ lb.}}{0.61\text{ in}} = 16.4\text{ lb/in}$$

Now we set the product-cushion system into motion by depressing the product 2 inches downward from the rest position and releasing it. Since this is a hypothetical product-cushion system, we will assume that there is no friction or damping so that the product or mass will vibrate indefinitely between ±2 inches.

What is the maximum velocity achieved by the product during the vibration? v_{max} may be calculated from Equation 3.18:

$$v_{max} = A\sqrt{\frac{kg}{W}} = 2\text{ in}\sqrt{\frac{(16.4\text{ lb/in})\,(386.4\text{ in/sec}^2)}{10\text{ lb}}}$$

$$= 50.3\text{ inches/second}$$

or, from Equation 3.19:

$$v_{max} = A\,(2\pi\,f_n) = 2\text{ in.}\,(2\pi\cdot 4\text{ cycles/sec}) = 50.3\text{ in/sec}$$

What is the maximum acceleration experienced by the product? From equation 3.20:

$$a_{max} = A \quad (\frac{kg}{W}) \quad = 2 \text{ in} \quad (\frac{16.4 \text{ lb/in} \quad 386.4 \text{ in/sec}^2}{10 \text{ lb}})$$

$$= 1267 \text{ in/sec}^2 = 3.28g$$

or, from Equation 3.21:

$$a_{max} = A\,(2\pi\, f_n)^2 = 2 \text{ in.}\,(2\pi \quad 4 \text{ cycles/sec})^2 = 1263 \text{ in/sec}^2$$

As our sample system vibrates at 4 cycles per second, the product reaches a maximum velocity of about 50 in/sec (2.84 miles per hour) at the midpoint of the motion. At the top and bottom extremes of the motion the acceleration reaches 3.28 g's.

As another example, let's look at the motion of a trailer truck body traveling the highway. The vibration of the truck body may also be approximated by the sinusoidal motion of a linear spring-mass system. Think of the truck body and the lading as the mass, and the springs of the suspension system as a linear spring. These springs are really not linear, but the usable part of the springs when working together are very close to linear. An external force sets this huge spring-mass system in motion when the truck goes over a pothole, expansion joint or other road surface discontinuity. Let's say that the weight of the trailer body is 20,000 lb. and it is loaded with 40,000 lb. of cargo. This weight deflects the springs 2 inches, so we may calculate an effective spring constant using Equation 3.11:

$$k = \frac{W}{\delta_{st}} = \frac{20{,}000 + 40{,}000 \text{ lb.}}{2 \text{ in.}} = 30{,}000 \text{ lb/in.}$$

If the truck should hit a pothole, the trailer will vibrate at (from Equation 3.15):

$$f_n = \frac{1}{2\pi}\sqrt{\frac{kg}{W}}$$

$$= \frac{1}{2\pi}\sqrt{\frac{(30{,}000 \text{ lb/in})\,(386.4 \text{ in/sec}^2)}{60{,}000 \text{ lb}}}$$

$$= 2.2 \text{ Hz}$$

Now suppose driving alongside the truck trailer vibrating at 2.2 cycles per second, we estimate the total vertical travel of the truck body during vibration at 3 inches (the double amplitude). With this information we may calculate the acceleration level at the trailer's floor. Using Equation 3.21:

$$a_{max} = A\,(2\pi\, f_n)^2, \text{ where } A = \frac{3 \text{ in.}}{2} \text{ and } f_n = 2.2 \text{ Hz}$$

$$a_{max} = 1.5 \text{ in } (2\pi \ 2.2)^2 = 286.6 \text{ in/sec}^2$$

or

$$\frac{286.6 \text{ in/sec}^2}{386.4 \text{ in/sec}^2 \text{ g}} = 0.74 \text{ g's}$$

This vibration of the truck bed is an external force which sets the lading into motion. The effects on the lading will be discussed in the next chapter.

3.5 Combination of Springs and Cushions

Before we leave simple vibration and move on to more complex motion, we need to develop a tool to deal with the combinations of springs and cushions with different vibration characteristics.[7]

Figure 3.11 shows two springs (or cushions) acting in **parallel** to support a weight, W.

The spring constants for each spring are known and need not be equal ($k_1 \neq k_2$). We would like to have an expression for a single equivalent spring constant, a "k_e" for an equivalent spring which, when substituted for the two springs shown, would behave exactly as the combinations does.

We expect k_e to be greater than either k_1 or k_2. Let's assume that the weight displaces each spring a distance of x downward. Therefore, spring one is exerting a force of $F_1 = k_2x$ at equilibrium and spring two is exerting a force of $F_2 = k_2x$.

$$F_1 + F_2 = W \qquad 3.22$$

The upward force of the springs exactly counteracts the downward force of the weight. The single force exerted by the equivalent spring must therefore equal the sum of the upward forces,

$$F_e = F_1 + F_2 \qquad 3.23$$

Substituting the equivalent values in 3.23,

$$k_ex = k_1x + k_2x$$

and factoring out x, leaves us with the expression

$$k_e = k_1 + k_2$$

Therefore, the equivalent spring constant for springs acting in parallel is simply the sum of the individual spring constants.

[7]Because we intend to apply our findings from the examination of springs to approximate the behavior of packaging cushions, which move only slightly under static and dynamic conditions, we will ignore the complications induced by an asymmetrical center of gravity, non-parallel spring motion, etc. This extremely simplistic approach may be used to estimate the behavior of combined packaging cushions.

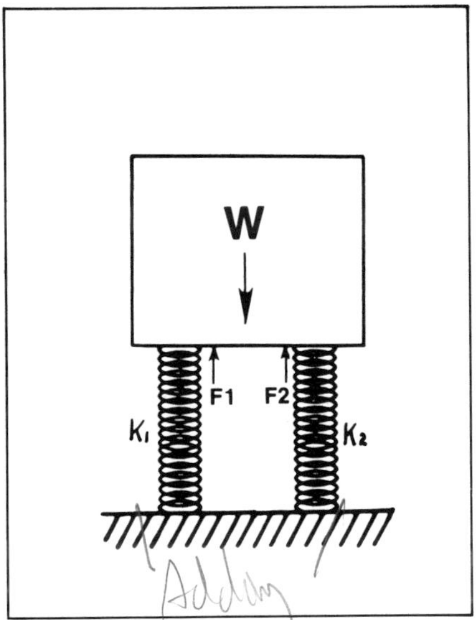

Figure 3.11. Two Springs Acting In Parallel.

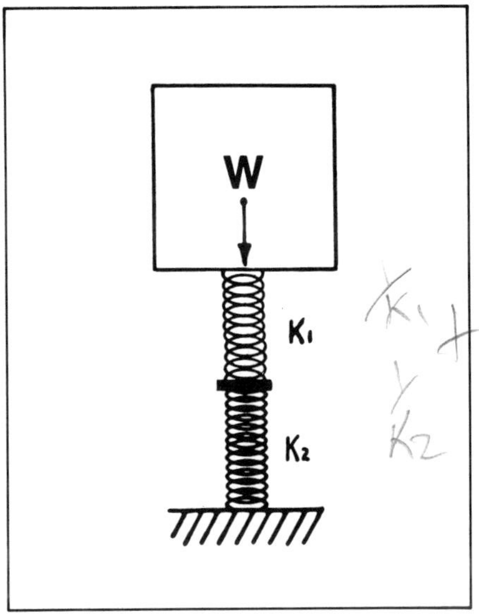

Figure 3.12. Two Springs Acting In Series.

Let's now examine the effect of putting springs or cushions in series, as shown in Figure 3.12. Again the system is compressed a distance of x, but now x is the sum of the compression of spring one and the compression of spring two:

$$x = x_1 + x_2 \tag{3.24}$$

Note that x is the static deflection of the system, so

$$x = \frac{W}{k_e}$$

and

$$x_1 = \frac{W}{k_1}, \quad x_2 = \frac{W}{k_2}$$

Substituting in Equation 3.24, we have:

$$\frac{W}{k_e} = \frac{W}{k_1} + \frac{W}{k_2}$$

Factoring out W leaves us with the expression we want:

$$\frac{1}{k_e} = \frac{1}{k_1} + \frac{1}{k_2} \tag{3.25}$$

Several examples will show how the relationships in Equations 3.24 and 3.25 may be used.

Suppose we have five springs all of equal length mounted side-by-side. Two of the springs have a k = 300 lb/in, the remaining three, k = 200 lb/in. A 100 lb. weight is placed on the springs. What is the static deflection of the system, and at what frequency will it vibrate when set into motion? To answer the question we must compute the equivalent spring constant from Equation 3.24 (because the springs are in **parallel**):

$$k_e = 300 + 300 + 200 + 200 + 200 = 1200 \text{ lb/in}$$

so

$$\delta_{st} = \frac{W}{k_e} = \frac{100 \text{ lb}}{1200 \text{ lb/in}} = 0.083 \text{ in}$$

and

$$f_n = 3.13\sqrt{\frac{1}{.083}} = 10.8 \text{ Hz}$$

Another example will show how the **series** relationship may be used. Imagine a piece of cushion measuring 10 in. x 12 in. x 3 in. which has been measured to have a spring constant of 400 lb/in. (For simplicity, we will assume that this cushion behaves like a linear spring.) Now we shave off the top one inch of the cushion. What is the new spring constant? To attack this problem, think of the original cushion to be three 10 in. x 12 in. x 1 in. thick cushions stacked on top of one another, as shown in Figure 3.13.

The equivalent spring constant for the three cushions in series is known to be 400 lb/in, so from Equation 3.25:

$$\frac{1}{400 \text{ lb/in}} = \frac{1}{k_1} + \frac{1}{k_2} + \frac{1}{k_3}$$

where k_1, k_2 and k_3 are the spring constants of each of the one inch thick cushions. Because the three pieces are identical, $k_1 = k_2 = k_3$,

$$\frac{1}{400 \text{ lb/in}} = \frac{3}{k_1}$$

and $k_1 = 3 \cdot 400$ lb/in $= 1200$ lb/in. Shaving one inch off the cushion leaves us with two one inch pieces for which k = 1200 lb/in, so the remaining 10 in. x 12 in. x 2 in. cushion has a spring constant of:

$$\frac{1}{k_e} = \frac{1}{k_1} + \frac{1}{k_2} = \frac{1}{1200} + \frac{1}{1200},$$

$$\frac{1}{k_e} = \frac{2}{1200} = \frac{1}{600}, \quad k_e = 600 \text{ lb/in.}$$

One last example will demonstrate how we may use the series and parallel concepts in combination to analyze a more complex cushion.

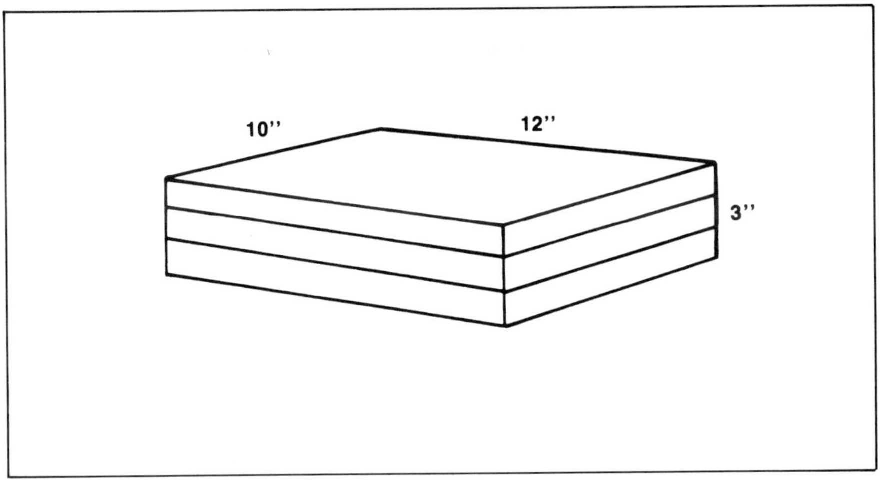

Figure 3.13. A 10 x 12 x 3 Inch Cushion.

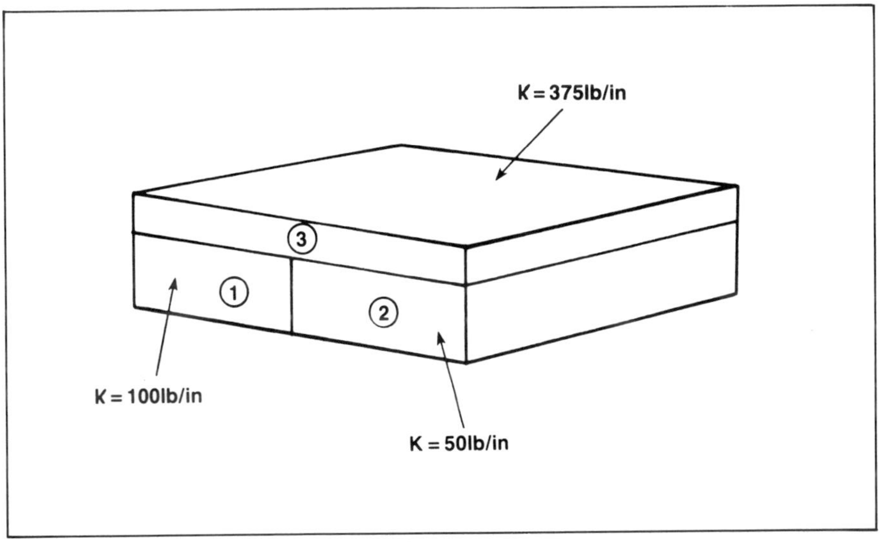

Figure 3.14. Three Cushions Acting Together.

Suppose we have three pieces of cushion material arranged as shown in Figure 3.14. What is the equivalent spring constant of the system?

To begin, we determine the effective spring constant of the lower layer where cushions 1 and 2 are operating in parallel:

$$k = 100 \text{ lb/in} + 50 \text{ lb/in} = 150 \text{ lb/in}$$

This lower layer is operating in series with cushion 3, so:

$$\frac{1}{k_e} = \frac{1}{150} + \frac{1}{375} = 0.0093$$

$$k_e = \frac{1}{.0093} = 107 \text{ lb/in.}$$

3.6 The Spring Constant and the Modulus of Elasticity

The modulus of elasticity, E, is a parameter that relates the stress and strain of a cushion material under a load. That is, E is the ratio of the force applied per unit area to the fractional change in thickness.[8]

More importantly, E is related to the spring constant:

$$k = E \frac{\text{area}}{\text{thickness}} \qquad 3.26$$

E, then, is a useful intermediate variable in solutions of cushion problems.

Let's take a slab of "linear" cushion material with an E measured to be 25 lb/in^2. If we cut a piece measuring 5 in. x 5 in. x 2 in. from this slab and use the 5 x 5 side as our bearing area, what is k for the cushion? Equation 3.25 may be used directly:

$$k = (25 \text{ lb/in}^2) \left(\frac{5 \text{ in. x } 5 \text{ in.}}{2 \text{ in.}}\right) = 312.5 \text{ lb/in.}$$

Likewise, we may use the above relationship to determine the result of changing the cushion's orientation. Suppose our cushion were turned on edge, and a 5 x 2 side used as the bearing area. Again:

$$k = (25 \text{ lb/in}^2) \left(\frac{5 \text{ in. x } 2 \text{ in.}}{5 \text{ in.}}\right) = 50 \text{ lb/in.}$$

[8] E = Force/Area ÷ Change in Thickness/Thickness.

PROBLEM SET, CHAPTER 3

1. A spring which is three inches long when not carrying a load deflects one inch when a 12 pound load is placed on it. What is the spring constant?

2. A spring having a free height of 4.25 inches is only 3.15 inches high when it supports a load of five pounds. What is the spring constant?

3. A spring deflects 0.55 inches when a twenty pound load is placed on it. What is the spring constant?

4. Compute the spring constant for a spring which deflects 2.27 inches when a 13.5 pound load is placed on it.

5. How much does the spring of Problem 1 deflect when a two pound load is placed on it?

6. Compute the deflection when the spring of Problem 4 is used to support a 15.8 pound load.

7. A rectangular object weighing 50 pounds is supported on four springs (one at each corner). All of the springs are the same and have a spring constant of 20 lbs/inch. How much is each spring deflected?

8. The springs in Problems 1 and 2 are placed in series (one on top of the other). What is the combined or effective spring constant?

9. The springs in Problems 1 and 2 are placed in parallel. What is the combined or effective spring constant?

10. Calculate the natural frequencies of the spring mass systems described in Problems 1 and 2.

11. Calculate the natural frequency of a spring mass system in which the springs are deflected 0.35 inches under the load which they support.

12. Find the natural frequency of a spring mass system where the static deflection is 3 inches.

13. A spring mass system is found to have a static deflection of 0.015 inches. What is the natural frequency of the system?

14. If a spring mass system has a natural frequency of 25 Hz, calculate the static deflection.

15. A cushion with k = 200 lb/in is placed on top of another cushion with k = 100 lb/in. A 120 lb. weight is then placed on top of these. What is f_n?

16. Solve the following:

A. A freely vibrating spring mass system has a spring (k = 30 lbs/inch) and a weight (2 lbs). Find the maximum acceleration if the maximum displacement is 1/2 inch.

B. Use the same spring given in A but with a weight of 6 lbs. Find the maximum acceleration with an initial displacement of 1/2 inch.

C. Repeat A using the same spring and displacement with a weight of 1 lb.

D. Repeat A using a 3 lb. weight with a spring of 200 lbs/inch. Displacement remains at 1/2 inch.

E. Find the acceleration in D when the initial displacement is 1.0 inch.

17. A vibration table with a fixed amplitude of 0.75 inches is required by a test procedure to operate at an acceleration of 2 g. At what frequency must the machine be operated?

18. A spring, 2.85 inches long when not carrying a load, deflects 0.67 inches when an 8.2 lb. weight is placed on it.

A. What is the spring constant k?

B. What is the natural frequency, f_n, of this spring mass system?

19. A vibrating spring-mass system has an amplitude of A = 1.7 inches, a spring constant k = 44 lbs/in. and a mass of weight 8 lbs. What is the maximum velocity of the mass expressed in inches/second? What is the maximum acceleration of the mass, expressed in inches/second^2?

20. A 10 lb. weight is placed upon a piece of linear cushion. The cushion is observed to compress 0.20 in.

A. What is the linear spring constant of this cushion?

B. At what frequency would this spring mass system vibrate if it were allowed to vibrate freely?

C. A 3 lb. weight and a 9 lb. weight are placed on top of the 10 lb. weight which sits on the cushion. What is the static deflection of the cushion?

D. What is the natural frequency of the system described in C?

21. A linear cushion measures 3 inches when not under a load. A 30 lb. weight is placed on the cushion. The cushion is then measured and found to be compressed to 2.73 inches.

   A. What is the linear spring constant of this cushion?

      Three layers of this cushion (i.e. 9 inches of unloaded cushion) are placed under a 5.55 lb. product.

   B. What is the equivalent spring constant of this cushion system?

   C. What is the natural frequency of this product-cushion system?

22. A cushion measuring 10 in. x 8 in. x 2 in. has a spring constant of k = 100 lb/in. What is the modulus of elasticity, E , for the cushion material? If the same cushion is now placed on end (with the 10 inch side vertical), what is the new spring constant?

23. The cushion in Problem 22 bottoms out under a load of 60 lb. What is its "working length"? When placed on end (10 inch side vertical), what load causes it to bottom out?

24. A piece of cushioning material measuring 10 in. x 15 in. x 3 in. has a spring constant of k = 100 lb/in. A 50 lb. product measuring 5 in. x 8 in. x 4 in. is placed on top of the cushion. Find the natural frequency of this spring mass system.

25. Write the equation of motion, y(t), with y as the vertical displacement and t as time for an object which executes oscillatory motion with a period of 400 ms and a peak acceleration of 0.7 g.

4. FORCED VIBRATION AND PRODUCT SENSITIVITY

4.0 Forced Motion

In the last chapter we examined the situation where a spring-mass system representing a product on a cushion or a component within a product was excited once by an outside force. The force was removed and the system allowed to vibrate freely. Now we want to consider what happens when a vibration is imposed on the spring-mass system from outside the system, by the vibrating floor of a vehicle for example. Figure 4.1 illustrates this situation. Here we have a spring mass system attached to a much larger vibrating mass. The **input** vibration occurs at a known frequency, the **forcing frequency f_f,** with a measurable amplitude and a maximum acceleration. We know the natural frequency, f_n, of the spring mass system; what we are interested in is the frequency of the system's vibration, its amplitude, and maximum acceleration resulting from the input or forcing motion.

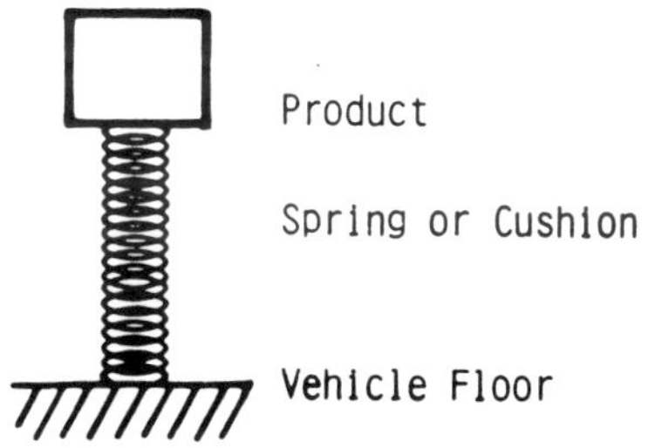

Figure 4.1. A Model of the Product, Cushion and Vehicle Combination.

From empirical investigation and mathematical analysis, three general statements relating the input and output vibrations may be made:

1. The output vibration will occur at the same frequency as the forcing vibration, not at the natural frequency of vibration[1].

2. The output amplitude of vibration of the spring mass system is directly related to the input amplitude of the forcing vibration by a calculable number, the magnification factor:

   Output Amplitude = Input Amplitude x Magnification Factor

   This factor will be discussed in the next two sections.

[1]There may be other frequencies present in the vibration response to the forcing vibration, but the amplitudes and accelerations associated with these frequencies are very much smaller than those of the vibration at the forced frequency, f_f.

3. The maximum acceleration experienced by the spring mass system is directly related to the maximum acceleration associated with the forcing vibration by the same magnification factor:

Maximum Output Acceleration =
Maximum Input Acceleration x Magnification Factor

The **magnification factor, M,** is defined simply as:

$$M = \frac{\text{Vibration Output or Response}}{\text{Vibration Input or Drive}} \qquad 4.1$$

4.1 The Magnification Factor, A Mathematical Derivation[2]

Let's consider the forcing motion to be harmonic, and represented by the expression:

$$y(t) = B \sin (qt) \qquad 4.2$$

and the output motion represented by x(t), following the conventions outlined in Section 3.0. We have two situations, shown in Figure 4.2, which are of interest.

In the first (a), the spring mass system is attached to a stationary base. If the spring mass system is set in motion by a single application of an outside force, the equation of motion is:

$$\frac{d^2x}{dt^2} + p^2x = 0$$

which has a solution

$$x = A \sin (pt)$$

where

A = initial displacement

$$p = \sqrt{\frac{kg}{W}}$$

as discussed in Section 3.3.

The more interesting situation, Figure 4.2 (b), where the base is vibrating, has a basic equation of motion:

$$\frac{d^2x}{dt^2} + p^2x(t) = p^2y(t) \qquad 4.3$$

where y(t) is represented by Equation 4.2.

[2]This section may be omitted without any loss of understanding of the practical aspects of forced vibration. This derivation follows that given in Reference 4.

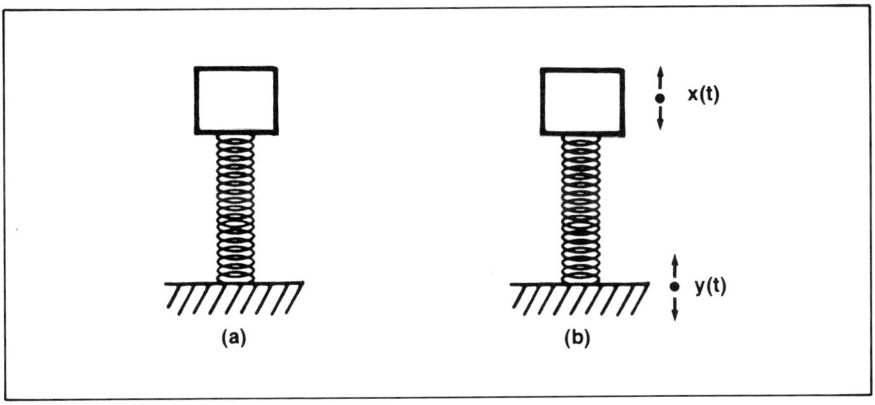

Figure 4.2. A Spring-Mass System on a Stationary Base and a Vibrating Base.

Let $q = 2\pi f_f$, where f_f is the forcing frequency, and $p2 = \frac{kg}{W}$ as before, and Equation 4.3 becomes:

$$\frac{d^2x}{dt^2} + p^2x = p^2 B \sin(qt) \qquad 4.4$$

We know from laboratory investigations that the spring mass system will vibrate predominately at the forcing frequency, so we may expect a solution to Equation 4.4 to take a form such as:

$$x(t) = D \sin(qt) \qquad 4.5$$

Accepting this, then:

$$\frac{dx}{dt} = D\, q \cos(qt)$$

$$\frac{d^2x}{dt^2} = -D\, q^2 \sin(qt) \qquad 4.6$$

Substituting Equation 4.6 into Equation 4.2, we get:

$$-Dq^2 \sin(qt) + p^2 D \sin(qt) = p^2 B \sin(qt)$$

and simplifying:

$$(p^2 - q^2) D \sin(qt) = p^2 B \sin(qt) \qquad 4.7$$

If Equation 4.7 is true for all values of t, then:

$$(p^2 - q^2) D = p^2 B$$

which may be rearranged:

$$\frac{D}{B} = \left(\frac{p^2}{p^2 - q^2}\right)$$

$$\frac{D}{B} = \left(\frac{1}{1 - \frac{q^2}{p^2}}\right)$$

where $q = 2\pi f_f$ and $p = 2\pi f_n$, so

$$\frac{D}{B} = \frac{1}{1 - \left(\frac{2\pi f_f}{2\pi f_n}\right)^2} = \frac{1}{1 - \left(\frac{f_f}{f_n}\right)^2} \qquad 4.8$$

B is the amplitude of the forcing or input vibration and D is the amplitude of the resulting or output vibration of the spring mass system. Equation 4.8 relates input to output motion:

$$\frac{\text{Output Amplitude}}{\text{Input Amplitude}} = \frac{1}{1 - \left(\frac{f_f}{f_n}\right)^2} \qquad 4.8a$$

which gives us an expression for the magnification factor:

$$M = \frac{1}{1 - \left(\frac{f_f}{f_n}\right)^2} \qquad 4.9$$

We may also generate an expression relating input and output acceleration. Recall Equation 3.9 where the maximum acceleration is given as a function of the amplitude and circular frequency:

$$a_{max} = Ap^2$$

From Equation 4.2, the maximum input acceleration is:

$$a_{maxI} = B\,q^2$$

From Equation 4.5, the maximum output acceleration is:

$$a_{max0} = D\,q^2$$

Returning to Equation 4.8, we may multiply both sides of the expression by:

$$\frac{q^2}{q^2} = 1 \;;$$

$$\frac{D\,q^2}{B\,q^2} = \left(\frac{q^2}{q^2}\right)\left(\frac{1}{1-\left(\frac{f_f}{f_n}\right)^2}\right) = \frac{1}{1-\left(\frac{f_f}{f_n}\right)^2}$$

The left hand side is now the ratio of the maximum output and input accelerations, while the right hand side remains the magnification factor:

$$\frac{a_{max0}}{a_{maxI}} = M \qquad 4.10$$

With the assumption that the input and output vibrations of a spring mass system may be expressed as single sinusoidal motions, an expression relating the amplitudes and accelerations of the motion has been derived.

4.2 Vibration Magnification

In the last section, a simple expression relating the parameters of the forcing or input motion to the forced or output motion was derived. The magnification factor may be expressed as:

$$M = \frac{\text{Output}}{\text{Input}} = \frac{1}{1-\left(\frac{f_f}{f_n}\right)^2} \qquad 4.11$$

where OUTPUT may be either the output (or forced) amplitude or the maximum output acceleration. Acceleration INPUT may be either the input (or forcing) amplitude or the maximum input acceleration.

f_f = forcing or input frequency

f_n = the spring mass system's natural frequency of vibration

Figure 4.3 contains a graph of the M factor as a function of the ratio of the forcing frequency to the natural frequency. Several points or regions on this curve are of interest.

On the far left hand side of the curve, where $\frac{f_f}{f_n}$ is very small, the value of M is very close to 1. This means that when the forcing frequency is very much smaller than the natural frequency, the output and input are nearly equal:

$f_f << f_n$,

$M \cong 1$, and

OUTPUT $\cong$ INPUT

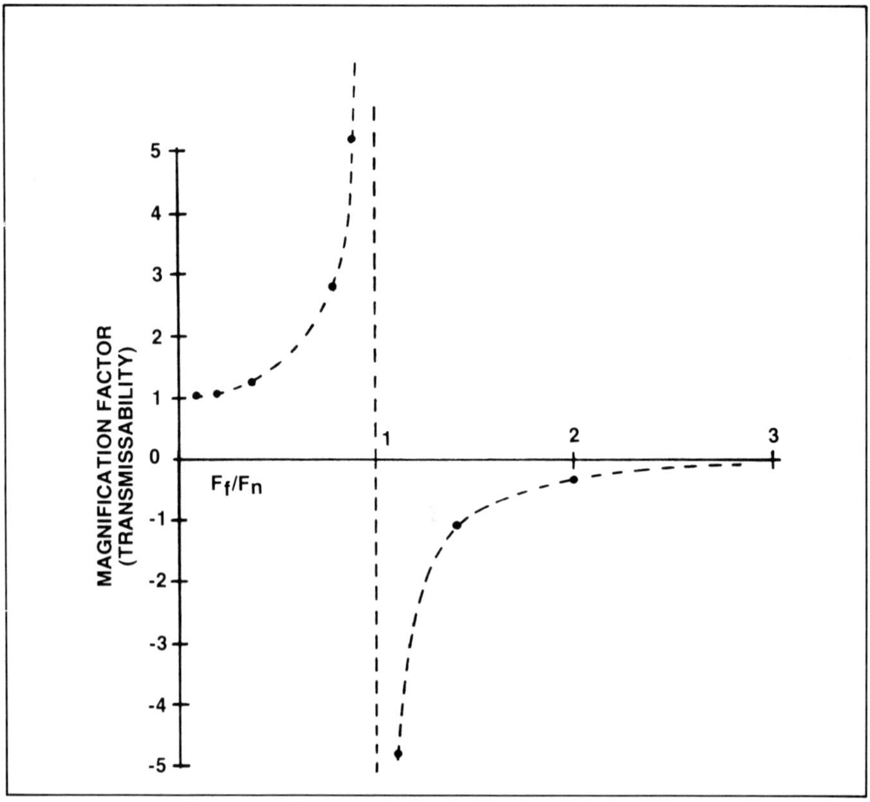

Figure 4.3. The Magnification Factor Shown as a Function of the f_f/f_n Ratio.

As the value of the forcing frequency approaches the value of the natural frequency, the value of M increases rapidly; approaching infinity as the ratio $\frac{f_f}{f_n}$ approaches 1:

$$\text{as } f_f \rightarrow f_n\text{, } M \rightarrow \infty$$

As this happens the output approaches infinity also. In reality, this means that the output is many times the value of the input.

At the point $\frac{f_f}{f_n} = 1$, where the forcing frequency exactly equals the natural frequency, the value of M becomes mathematically undefined:

$$M = \frac{1}{1 - \left(\frac{ff}{fn}\right)^2} = \frac{1}{1 - \left(\frac{1}{1}\right)^2} = \frac{1}{0}$$

This is **resonance,** and in reality means that a very small input vibration will cause extremely large output vibrations.

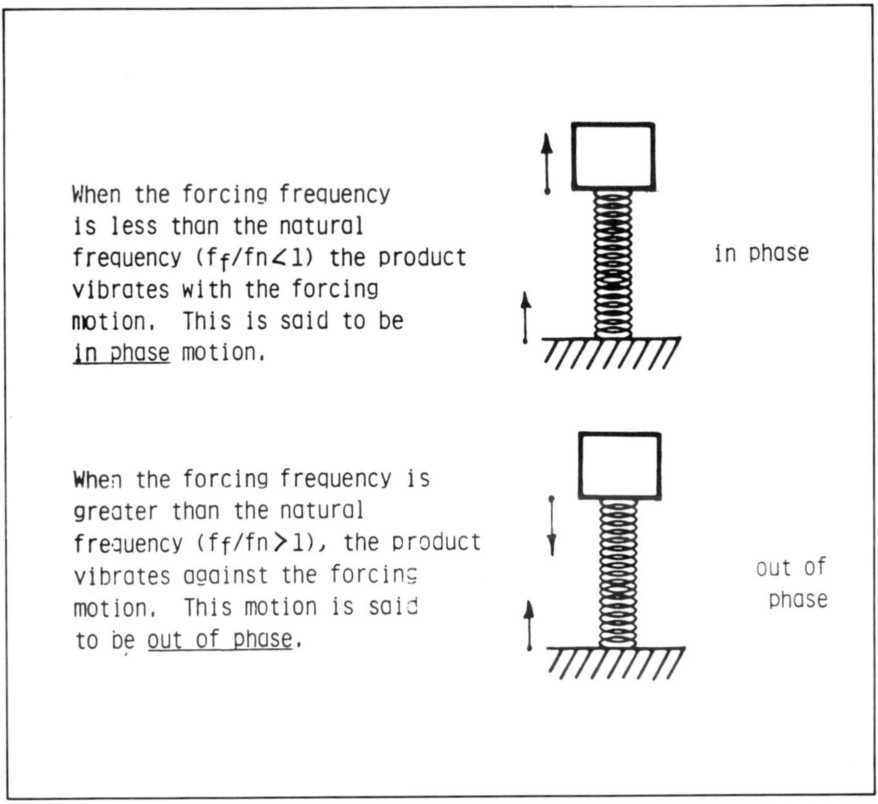

Figure 4.4. In Phase and Out of Phase Motion.

Between the points $\frac{f_f}{f_n} = 0$ and $\frac{f_f}{f_n} = 1$, the value of M is a **positive** number. A positive value of M is an indication that the forcing motion and the resulting forced motion are **in-phase.** As shown in Figure 4.4, for in-phase motion, as the base moves upward the spring mass system also moves upward. When the base moves downward, the spring mass follows.

For frequency ratios larger than one, $f_f > f_n$, the value of M is a **negative** number. This negative value indicates that the forcing and forced motions are **out of phase.** As shown in Figure 4.4, out of phase motion is movement such that the base and the spring mass system are always traveling in opposite directions (except for the end points of the motion where the velocity is zero).

Recognizing that in-phase motion occurs for points to the left of $\frac{f_f}{f_n} = 1$, and out of phase motion for points to the right, the magnification curve may be simplified to that shown in Figure 4.5. Here the absolute value of M ($|M|$) has been plotted against the frequency ratio. The information contained in the sign of the magnification factor has been preserved in the phase relationship. This curve, or similar ones, are often referred to in packaging literature as response or transmissibility curves.

As we move to the right from $\frac{f_f}{f_n} = 1$ on the curve in Figure 4.5, the value of $|M|$ decreases. This means that as the forcing frequency becomes increasingly larger than the natural frequency, the magnification value progressively decreases.

Notice that at a value slightly less than $\frac{f_f}{f_n} = 1.5$ the value of $|M|$ falls below one. This point is often referred to as the **isolation point.** All values of $|M|$ to the right of this point are less than one; the output is always than the input for ratios in this region.

There is only one isolation point. For values of $\frac{f_f}{f_n}$ less than one, $|M|$ is **always** greater than one. The isolation point occurs when $M = -1$ and the value of $\frac{f_f}{f_n}$ is:

$$-1 = \frac{1}{1 - (\frac{f_f}{f_n})^2}$$

$$1 - (\frac{f_f}{f_n})^2 = -1,$$

$$(\frac{f_f}{f_n})^2 = 2,$$

$$\frac{f_f}{f_n} = \sqrt{2} = 1.414 \qquad 4.12$$

For situations where

$$\frac{f_f}{f_n} > \sqrt{2},$$

$$\frac{\text{OUTPUT}}{\text{INPUT}} < 1,$$

Forcing vibrations occur throughout the packaging distribution system. Warehousing and transportation by truck, rail, ship and air each produce characteristic forcing vibrations. Obviously, to avoid large magnification values our products and package systems should not have natural frequencies similar to these forcing frequencies.

4.3 Vibration Magnification: Sample Problems

Let's now apply some of the relationships developed in the last several sections to situations of interest to packagers. Suppose we have a product in a package which deflects its cushion 2.0 inches under normal one g circumstances. What happens when this product-package system is transported in a vehicle which has a forcing vibration at 2.5 Hz, with a maximum associated input acceleration of 0.1 g? That is, what is the maximum displacement of the product on its cushion and the maximum

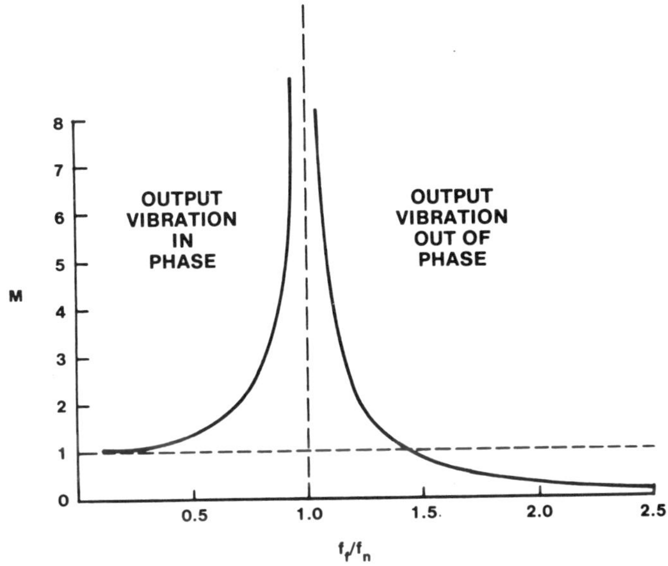

Figure 4.5. The Absolute Value of the Magnification Factor Shown as a Function of the Frequency Ratio

associated acceleration of the product as a result of the forcing vibration? We are given that:

$$\delta_{st} = 2.0 \text{ in. ,}$$

$$f_f = 2.5 \text{ Hz and}$$

$$a_{maxI} = 0.1 \text{ g}$$

First, we need to calculate the natural frequency of the product on its cushion from the information on the static deflection:

$$f_n = 3.13 \sqrt{\frac{1}{2.0}} = 2.21 \text{ Hz}$$

Notice that because the natural frequency is less than the forcing frequency, we expect the output motion to be out-of-phase. Now the magnification factor may be computed:

$$M = \frac{1}{1 - \left(\frac{f_f}{f_n}\right)^2} = \frac{1}{1 - \left(\frac{2.5}{2.21}\right)^2} = -3.58$$

(The minus sign tells us that indeed the input and output motions are out of phase.)

From Equation 4.10 or 4.11, we may relate the maximum input acceleration to the maximum output acceleration:

$$\frac{a_{max0}}{a_{maxI}} = M$$

Then:

$$a_{max0} = M \cdot a_{maxI}$$

$$= (-3.58)\ (0.1\ g) = -0.358\ g\text{'s}$$

The maximum output acceleration is roughly 0.36 g's, out of phase with the input motion.

The displacement motion of the product on its cushion may be determined in one of two equivalent ways. Just as the maximum acceleration of a mass on a freely vibrating spring is related to the maximum displacement amplitude and the frequency of vibration (Equation 3.21),

$$a_{max} = A(2\pi f_n)^2 \quad ,$$

the maximum acceleration of the product on the cushion vibrating at the forcing frequency is related to the maximum displacement, amplitude and the forcing frequency by the relation:

$$a_{max0} = A_0\ (2\pi f_f)^2 \qquad 4.13$$

where a_{max0} is the maximum **output** acceleration and A_0 is the maximum **output** (single amplitude) displacement describing the motion of the product on the cushion. Solving for A_0:

$$A_0 = \frac{a_{max0}}{(2\pi f_f)^2}$$

and substituting the known values,

$$A_0 = \frac{(-0.358\ g)\ (386.4\ \text{in/sec}^2 \cdot g)}{(2\pi\ \ 2.5\ \text{cycles/sec})^2} = -0.56\ \text{in.}$$

(Notice that we converted the acceleration in g's to in/sec^2.) The product under this forced vibration is displaced ±0.56 inches from the rest or static position, and the product's motion is out of phase with the forcing motion of the floor of the vehicle.

In another method of computing the displacement of the product on the cushion we first compute the displacement of the vibrating floor of the vehicle. As in Equation 4.13, we know the relationship between maximum displacement, maximum acceleration and vibration frequency:

$$a_{maxI} = A_I\ (2\pi f_f)^2 \qquad 4.14$$

where a_{maxI} is the maximum **input** acceleration and A_I is the maximum **input** (single amplitude) displacement. Again solving for A_I and substituting the known values, we have:

$$A_I = \frac{a_{maxI}}{(2\pi f_f)^2}$$

$$= \frac{(0.1 \text{ g})(386.4 \text{ in/sec}^2 \text{ g})}{(2\pi \ 2.5 \text{ cycles/sec})^2} = 0.157 \text{ in.}$$

Knowing that the maximum single amplitude displacement of the vehicle's floor is 0.157 inches, we may now use the relationship between input and output of Equation 4.8a:

$$\frac{A_0}{A_I} = M$$

Therefore:

$$A_0 = M \cdot A_I = (-3.58)(0.157 \text{ in.}) = -0.56 \text{ inches}$$

Again, the minus sign indicates that the input vibration of the vehicle's floor and the resulting output vibration of the product on its cushion are out of phase.

In summary, for this example we see an input vehicle vibration at 2.5 Hz with a maximum, single amplitude displacement of 0.157 inches and maximum acceleration of 0.1 g resulting in a product vibration on its cushion at 2.5 Hz with a maximum single amplitude displacement of 0.56 inches and maximum acceleration of 0.358 g.

Now let's examine a more complex example where we can use the simple relationships of magnification to estimate the effects of distribution vibration on a product-package system.

A product-package system is shown in Figure 4.6a. The product has a small internal component which is known to have a natural frequency of 3 Hz (this may be a small circuit board, transistor or similar mechanical/electrical part). When the product is placed in the package a cushion deflection of 0.27 inches is observed. This product-package system is to be transported in a truck with a 4 Hz vibration frequency and associated maximum acceleration of one-half g. (Incidently, this is a typical vibration found in trucks.) What will happen to this product in transport?

The model shown in Figure 4.6b may be used to simplify matters and help us formulate the problem. Here, the product is depicted as a mass, m_1, on a linear spring representing the cushion. Since the static deflection of the cushion is known, the natural frequency of the linear spring may be computed:

$$f_n = 3.13 \sqrt{\frac{1}{0.27}} = 6 \text{ Hz}$$

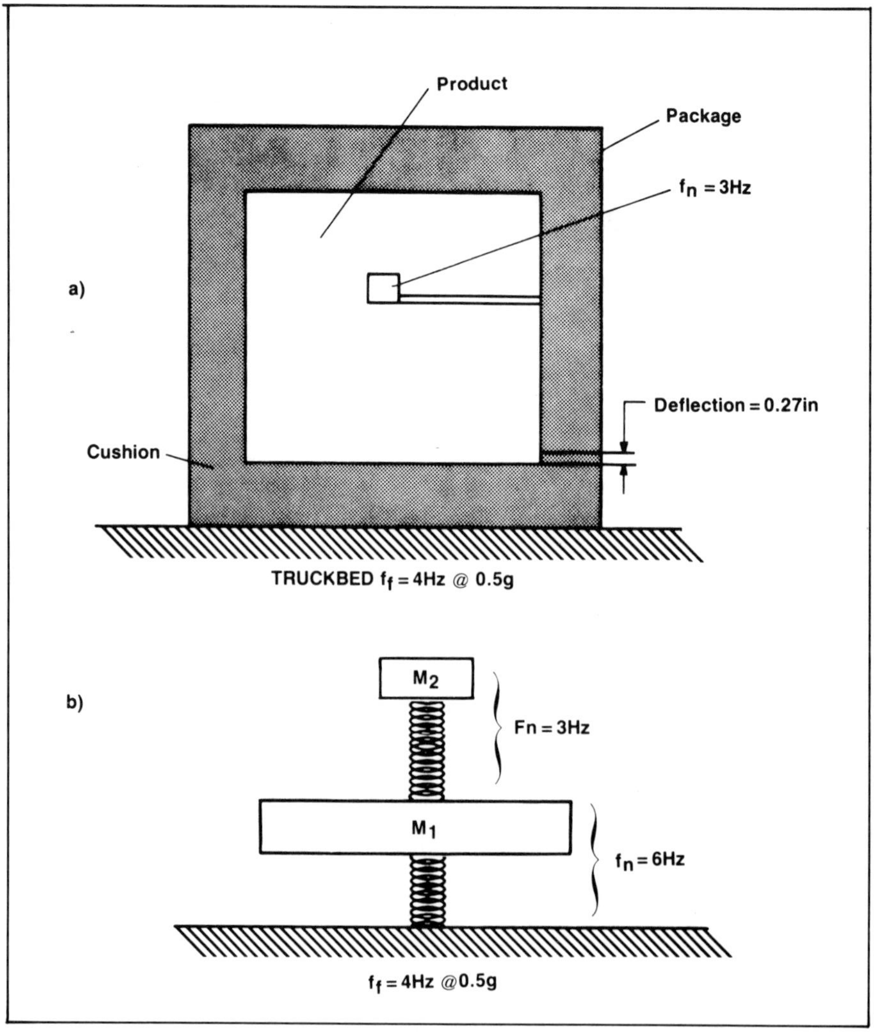

Figure 4.6. A Product-Package System.

The 3 Hz component is shown as mass, m_2, also on a linear spring. We will assume that the weight of the component is very small compared to that of the product, i.e. $m_2 << m_1$. This assumption and the substitution of the linear spring for the cushion will allow us to apply the previous methodology to this problem.

First, we will calculate the motion of the truck bed using Equation 4.14:

$$a_{maxI} = A_I (2\pi f_f)^2$$

$$A_I = \frac{a_{maxI}}{(2\pi f_f)^2} = \frac{(0.5 \text{ g})(386.4 \text{ in/sec}^2 \text{ g})}{(2\pi \ 4 \text{ cycles/sec})^2} = 0.31 \text{ inches}$$

The forcing motion is now completely characterized;

forcing frequency: 4 Hz

maximum acceleration: 0.5 g

maximum single amplitude displacement: 0.31 in.

The magnification[3] factor linking the truck bed and the product on its cushion may now be calculated.

$$M_{TP} = \frac{1}{1 - (\frac{f_f}{f_n})^2} = \frac{1}{1 - (\frac{4 \text{ Hz}}{6 \text{ Hz}})^2} = +1.8$$

The positive value of M_{TP} indicates that the forcing motion of the truck bed and the forced motion of the product on its cushion are in phase.

The maximum single amplitude displacement of the product (m_1) may be calculated using Equation 4.8a:

$$\frac{A_0}{A_I} = M_{TP}$$

$$A_0 = A_I \cdot M_{TP} = (0.31 \text{ in.})\,(1.8) = 0.56 \text{ inches}$$

Similarly, from Equation 4.10 the maximum acceleration of the product may be calculated:

$$\frac{a_{max0}}{a_{maxI}} = M_{TP}$$

$$a_{max0} = a_{maxI} \cdot M_{TP} = (0.5 \text{ g})\,(1.8) = 0.9 \text{ g}$$

The output motion at the product may now be characterized;

frequency of vibration: 4 Hz

maximum acceleration: 0.9 g

maximum single amplitude displacement: 0.56 inches

motion in phase with the truck bed

The next step is to estimate the response of the 3 Hz component to the motion of the product.

---

[3]The magnification factor is given the subscript "TP" to indicate the transfer of motion from the Truck to Product.

The magnification[4] factor linking the product and component may be calculated:

$$M_{PC} = \frac{1}{1 - (\frac{f_f}{f_n})^2} = \frac{1}{1 - (\frac{4\text{ Hz}}{3\text{ Hz}})^2} = -1.3$$

The negative value of M_{PC} indicates that the motion of the component is out of phase with that of the product on its cushion.

The maximum single amplitude displacement of the component may be calculated as before, using the displacement of the product as the input value:

$$\frac{A_0}{A_I} = M_{PC}$$

$$A_0 = A_I \cdot M_{PC} = (0.56)(-1.3) = -0.73 \text{ inches}$$

Likewise, the maximum acceleration of the component may be calculated using the acceleration of the product as the input value:

$$\frac{a_{max0}}{a_{maxI}} = M_{PC}$$

$$a_{max0} = a_{maxI} \cdot M_{PC} = (0.9\text{ g})(-1.3) = -1.17\text{ g}$$

The output motion at the component may be characterized:

frequency of vibration: 4 Hz

maximum acceleration: 1.17 g

maximum single amplitude displacement: 0.73 inches

motion out of phase with the product and truck bed

From this simple analysis we see that the input vibration from the truck bed has been significantly magnified in the motion of the internal component. Before shipping we need to know if the product can withstand long periods of 4 Hz vibration at 0.9 g, if the component can withstand long periods of 4 Hz vibration at 1.17 g, and if there is enough space within the product to accommodate the large vibrational displacements of the component. If the answer is no to any of these questions, then either the product or the package or both need to be redesigned before distribution.

---

[4]The subscript "PC" indicates Product to Component

PROBLEM SET, CHAPTER 4

1. If a product in a package deflects its cushion a distance of 0.50 inches, what happens if the package-product is transported on a vehicle which has a forcing vibration input of 0.10 g at 4.5 Hz. Find the magnitude of vibration of the item on the springs. Calculate the amplitude of the displacement of the package on its cushion.

2. What would happen if the forcing frequency in Problem 1 was 4.3 Hz? Compute M and the maximum acceleration.

3. What happens to M and the acceleration in Problem 1 if f_f = 7.5 Hz?

4. A truck moving at a constant speed is instrumented with accelerometers. One of the accelerometers gives a reading of 2.4 g's at 16 Hz. What is the amplitude of the motion?

5. If a product deflects its cushion 0.76 inches, what is the natural frequency of the product-package system? What is the maximum amplitude and acceleration of the product if the product-package is transported in a vehicle with a forcing vibration of 0.3 g at 4.1 Hz? Is the forced vibration of the product in phase or out of phase with the 4.1 Hz forcing vibration?

6. A spring mass system with a natural frequency f_n = 16 Hz is attached to a vibration table. The table is set to vibrate at 12 Hz, 0.5 g maximum acceleration:

   A. What is the amplitude of the table's motion in inches?

   B. What is the magnification factor M in this situation?

   C. What is the maximum displacement of the mass assuming 0 damping?

   D. What is the maximum acceleration of the mass assuming 0 damping?

   E. Is the motion of the mass in phase or out of phase with the motion of the table?

7. A product sits at rest upon a piece of linear, undamped cushioning material. The static deflection is measured to be 0.10 in. The product and its cushion are placed on a vibration table which is set to vibrate at 8 Hz with a single amplitude maximum displacement of 0.5 inch.

   A. What is the resulting single amplitude maximum displacement of the product?

   B. What is the maximum acceleration experienced by the product during the vibration?

8. The product of Problem 21, Chapter 3, with its 9 inch cushion is to be transported in a truck known to have a forcing frequency of 7 Hz, with a maximum acceleration of 0.5 g.

   A. What is the maximum acceleration of the product during transport? Is this in or out of phase with the truck bed?

   B. What is the maximum displacement of the product from its rest position during transport?

9. A freely vibrating spring mass system is observed to have an amplitude (0-peak) of 1.0 inch, with a maximum acceleration of 1.0 g.

   A. What is the natural frequency of this spring mass system?

   This spring mass system is placed on a vibration table set to vibrate at 4 Hz, with a maximum (0-peak) displacement of 0.6 inch.

   B. What is the maximum acceleration of the vibration table?

   C. What is the maximum acceleration of the mass due to the forcing vibration of the table?

   D. Is the mass in or out of phase with the table?

10. A product and its cushion are placed on a vibration table which is set to vibrate at 18 Hz and 0.5 g's. The product responds with 2.0 g's.

   A. What is the zero-to-peak (single amplitude) displacement of the vibration table?

   B. At what frequency does the product cushion system vibrate under these conditions?

   C. Not knowing whether the product vibrations are in phase or out of phase with the table, what are the two possible natural frequencies of the product cushion system?

5. DAMPED VIBRATIONS, VIBRATIONS IN DISTRIBUTION, VIBRATION TESTING AND RANDOM VIBRATION

5.0 Damped Vibration

A vibrating object if undisturbed will eventually stop vibrating. A tuning fork is a good example; bang it on your elbow and it will vibrate vigorously for a few seconds, but within a minute all vibration will have stopped. The energy that was put into the system causing the vibration is eventually dissipated through friction, a phenomenon referred to as **damping.**[1]

A spring-mass system is said to have a damping coefficient, C, which describes the system's frictional dissipation of vibration energy.[2] This coefficient is measured relative to a reference value, C_c, the coefficient of **critical damping:**

$$C_c = 2\sqrt{\frac{kw}{g}} \qquad 5.1$$

The ratio of the system's actual damping coefficient to the critical value is often given as ξ :

$$\xi = \frac{C}{C_c}$$, the **fraction of critical damping**

When the spring-mass system is critically damped ($\xi = 1$), a displacement of the mass will not set the system into vibration. Instead, the mass will quickly return to its equilibrium position without further motion, as shown in Figure 5.1. In the overdamped situation ($\xi > 1$) the mass will also return to the rest or equilibrium position, without "overshooting," but more slowly. The critical damping coefficient is in fact based on the situation where the damping produces the most rapid return to equilibrium without further vibration.

Underdamping is a more common situation, at least in terms of our everyday experience with vibration. An underdamped spring-mass system ($\xi < 1$), after being set in motion, will vibrate at a constant frequency but with decreasing amplitude, as shown in Figure 5.2.

The natural frequency of this free vibration is given by the expression (13),

$$f_n' = \frac{1}{2\pi}\sqrt{\frac{kg}{W}(1 - \xi^2)} \qquad 5.2$$

---

[1]This discussion is based on a consideration of viscous damping. **Viscous damping** is damping in which the force resisting vibration is proportional to the velocity of vibration.

[2]This discussion follows the more technical one given in Reference 13, a brief but excellent explanation.

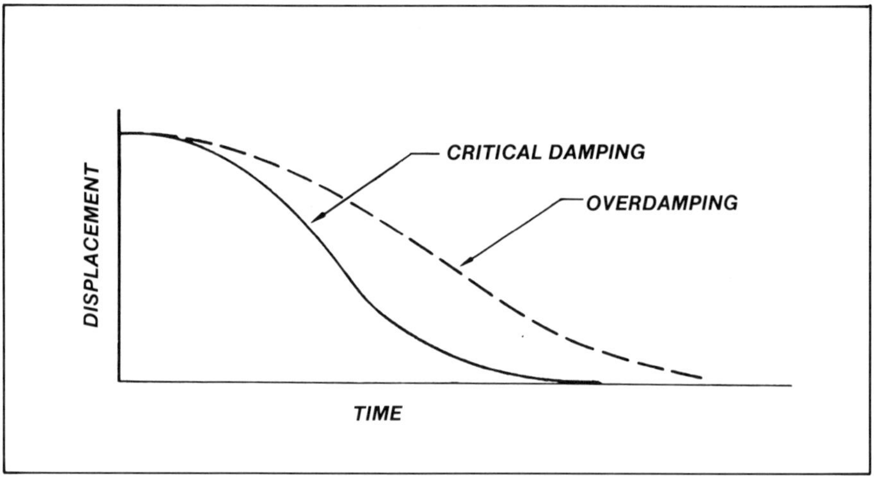

Figure 5.1. Return of a Damped Spring-Mass System to the Equilibrium Position. (13)

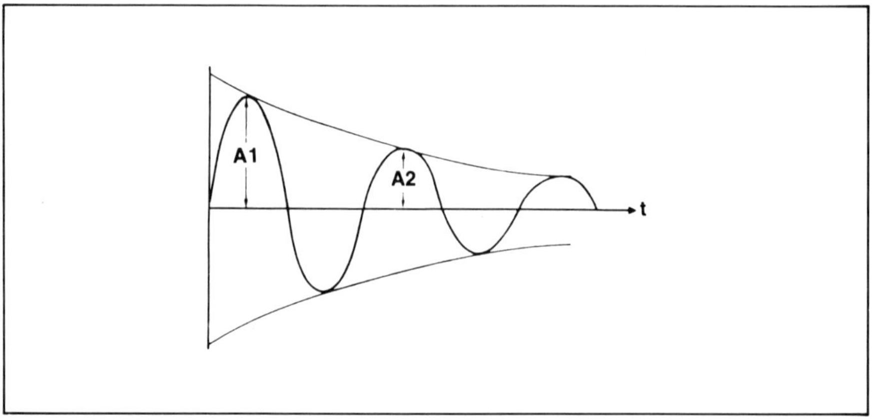

Figure 5.2. Motion of an Underdamped Spring-Mass System. (13)

The frequency of the damped vibration is very similar to the natural frequency of the undamped system, given by Equation 3.15. An inspection of the two equations reveals that

$$f_n' = f_n \sqrt{(1 - \xi^2)}$$

The difference between the two frequencies is therefore very small for most underdamped systems, so small that f_n is often used as the resonance frequency of the underdamped system.

The decrease in displacement amplitude from one cycle to the next due to the dissipation of energy through damping may be calculated from Equation 5.3 a or b:

$$\ln\left(\frac{A_1}{A_2}\right) = \frac{2\pi\xi}{\sqrt{1-\xi^2}} \qquad 5.3a$$

or

$$\frac{A_1}{A_2} = e^{\left(\frac{2\pi\xi}{\sqrt{1-\xi^2}}\right)} \qquad 5.3b^3$$

For a system with a known damping coefficient, Equations 5.1 and 5.3 allow us to calculate the rate at which a vibration will die out. Suppose, for example, that we have calculated the fraction of critical damping from Equation 5.1:

$$\xi = 0.1$$

From Equation 5.3a, the natural log of the ratio of the displacement amplitudes for any two successive vibrations is:

$$\ln\left(\frac{A_1}{A_2}\right) = \frac{2\pi\ (0.1)}{\sqrt{1-(0.1)^2}} = 0.63$$

$$\frac{A_1}{A_2} = e^{0.63} = 1.88$$

$$A_2 = \frac{A_1}{1.88} = 0.53\ A_1$$

This tells us that the maximum displacement during any one cycle is 53% of the maximum displacement during the previous cycle. If the displacement during the first cycle was one unit, the displacements during the succeeding vibrations would be as follows:

Cycle 1	1.00 unit
Cycle 2	0.53 unit
Cycle 3	0.28 unit
Cycle 4	0.15 unit
Cycle 5	0.08 unit
Cycle 6	0.04 unit
Cycle 7	0.02 unit
Cycle 8	0.01 unit

If the natural frequency[4] of vibration for this example was 1 cycle/sec (1 Hz), it would take eight seconds for the amplitude of the motion to decay from 1.00 to 0.01 unit, a factor of 100.

---

[3]e is the base for natural logarithms.

[4]The natural frequency calculated for the undamped system (Equation 3.15) would be 1.005 Hz compared to the damped system's 1.000 Hz natural frequency, an insignificant difference.

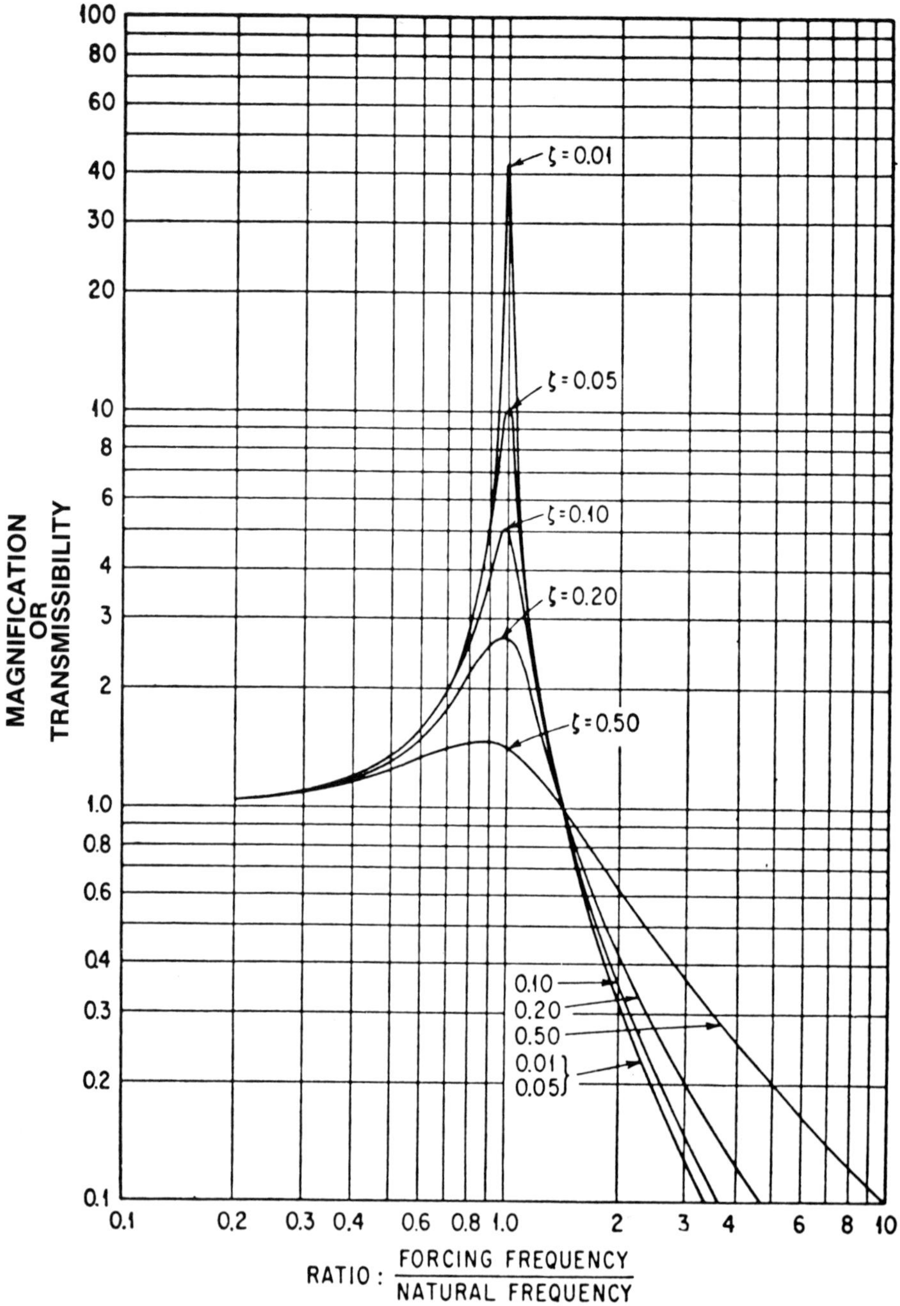

Figure 5.3. Magnification or Transmissibility of a Damped System. (6)

As one might expect, **damping also influences** the result of **forced motion.** Figure 5.3 contains a plot of the magnification factor or transmissibility as a function of the ratio of the forcing frequency to natural frequency for several damped, spring-mass systems. Notice that the magnification factor decreases as the fraction of critical damping increases. In packaging, we tend to use the easily calculated value for M presented in Chapter 4 to **estimate** vibration motion transfer, knowing that the actual value will be somewhat less due to the presence of damping. This conservative approach may produce overpackaging if prototype product-package systems are not refined and tested in the laboratory.

The transmissibility, T, for a linear spring-mass system having one degree of freedom may be calculated from the equation (6):

$$T = \sqrt{\frac{1 + [2\xi (f_f/f_n)]^2}{[1 - (f_f/f_n)^2]^2 + [2\xi (f_f/f_n)]^2}}$$

For a damped system with a known value for ξ, T may be used in place of M to calculate the resultant displacements and accelerations.

Damping also complicates the phase relationship between the forcing vibration and the output motion. In the undamped case discussed in Chapter four we saw that the forcing and resultant vibrations are in phase when f_f/f_n is less than one, and out of phase when f_f/f_n is greater than one; **phase angles** of 0° and 180°, respectively. Figure 5.4 shows the phase angles for several damped systems. Again, the undamped system ($\xi = 0$) is used to approximate the real situation for packaging purposes.

Excellent, technical discussions of damped vibration are contained in the Shock and Vibration Handbook (Reference 6) and Thomson's Theory of Vibrations with Applications (Reference 16).

5.1 Vibration Sensitivity

The vibration sensitivity of a product may be defined as those input vibration frequencies of sufficient amplitude occurring in the distribution environment which can cause the resonance and failure of the product or components within the product. Information on the vibration sensitivity may be used either to properly design a package system that will minimize the product's response to the input vibrations in distribution or to redesign the product.

The following three step method may be used to determine vibration sensitivity:

1. Identify the natural frequencies of the product.

2. Identify the natural frequencies which occur in the distribution environment.

3. Determine whether failure occurs when the product is vibrated at the sensitive frequencies at the amplitudes encountered in the distribution environment.

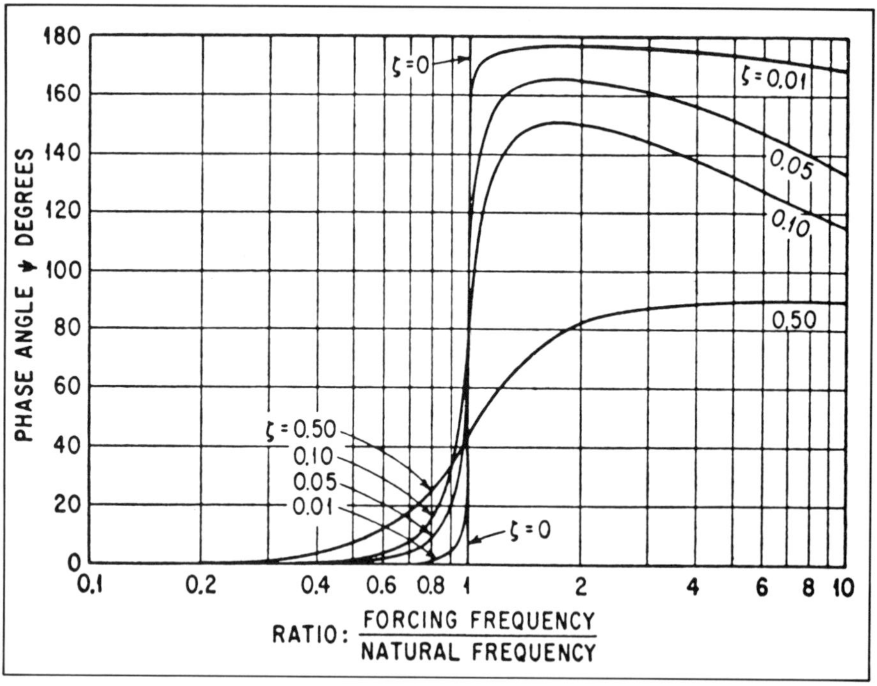

Figure 5.4. Phase Angles for Damped Systems. (6)

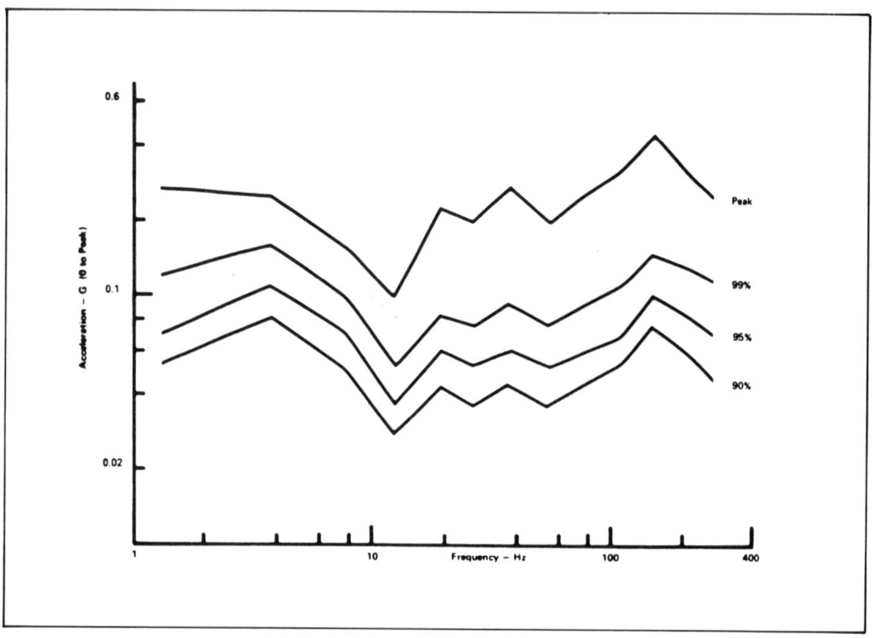

Figure 5.5. Vibration in Distribution: Frequencies, Acceleration Levels and Probability of Occurrence. (22)

The natural frequencies of a product are usually identified using a Vibration Table[5] capable of producing sinusoidal motion. The product is securely attached to the table, and the system is vibrated at frequencies between 3 Hz and 100 Hz with an associated maximum acceleration of between 0.1 to 0.5 g (zero to peak amplitude). A constant 0.5 g's is commonly employed throughout the frequency range sweep. The natural frequencies of the product are those which cause components to resonate, identifiable either visually or by the intense sounds of the vibration. The frequency sweep may be interrupted and the frequency kept constant to confirm the resonant frequency and the component affected.[6]

This procedure simply identifies the natural frequencies of a product in the range of 3 to 100 Hz. These natural frequencies may or may not be fragile points of the product.

Vibration frequencies encountered in the distribution environment are listed in Table 5.1, along with the maximum associated acceleration levels. Vibration in distribution is extremely complex, with wide variations in frequencies and amplitudes. These general data indicate some of the ranges that most often occur.[7]

Another manner of presenting vibration information is shown in Figure 5.5. Here frequencies, acceleration levels and probabilities of occurrence are interrelated. The important thing to emphasize here is that if any of the natural frequencies of the product match up with frequencies occurring in the proposed distribution system, careful product testing is needed.

TABLE 5.1
VIBRATIONS ENCOUNTERED IN DISTRIBUTION

Transport	Frequency		Maximum G Level
Rail Car	Suspension frequency	2 - 7 Hz	1/2 g
	Suspension, lateral	3/4 - 2 Hz	3/4 g
	Structural	50 - 70 Hz	1/4 g
	Roll	approx. 1 Hz	—
Truck	Suspension	2 - 7 Hz, 4 Hz	1/2 g
	Unsprung suspension	10 - 20 Hz	1/4 g
	Structural	50 - 100 Hz	1/4 g
Trucks on Flat Cars	Vertical	2 - 4.6 Hz	1 g
	Roll	0.7 - 3.1 Hz	10 g
Aircraft	Propeller	2 - 10 Hz	—
	Jet	100 - 200 Hz	—
Ships	Sea and Engines	10, 100 Hz	—

[5]Vibration tables are commercially available in many forms.

[6]This procedure is further outlined in ASTM Standard D 3580, "Vibration (Vertical Sinusoidal Motion) Test of Products." See Reference 21.

[7]The interested reader should consult Reference 15 for a more detailed review of vibration in distribution.

The purpose of product testing is to determine if a product is damaged when exposed to vibrations at its natural frequencies. The product is again securely attached to the vibration table which is set to vibrate at each of the product's natural frequencies, one frequency at a time with an appropriate, associated acceleration level. The duration of the test at each frequency is a matter of judgment; thirty minutes is usually a good starting point. The product is examined for damage after each vibration test. The frequencies which cause product failure are of concern. Either the product must be packaged to protect it from such vibrations or the product must be redesigned to eliminate or reduce resonance at frequencies occurring in the distribution environment.

5.2 Vibration of the Packaged Product[8]

Test methods have been designed to evaluate the performance of a packaging system when it is subjected to the vibration conditions existing in the distribution environment. If a packaging system fails these tests it should be redesigned. However, if it passes the tests there is no guarantee that the packaged product will survive the shipment.

There are three testing methods available depending on how the product is to be shipped:

1. The Repetitive Shock Test should be used if individual containers of products are shipped.

2. The Single Container Resonance Test is done to assess the container's ability to protect the product when the input vibration has the same frequency as that of the container with the product.

3. The Unitized Load or Vertical Stack Resonance Test should be done if containers are shipped in multiple unit, stacked loads.

These tests are briefly described below.

The **Repetitive Shock Test** is used for individually shipped packaged products which may be exposed to repetitive mechanical shocks on the truck bed. The test specimen is placed on the vibration table in its normal shipping orientation with restraining devices present to prevent the specimen from dropping off the table. The vibration table is started at 2 Hz with a fixed displacement of one-half inch and the frequency is gradually increased until the test specimen begins to bounce on the table or the acceleration of the table is 1.1 g (zero to peak). At this frequency the test is continued for thirty minutes and the specimen is then examined for damage to both the product and the package.

In the **Single Container Resonance Test** the test specimen is restrained on the vibration table in its normal shipping orientation to simulate a package in a unitized load or an individual container under other packages. The vibration table is started

---

[8]This method is further described in ASTM D 999, "Vibration Testing of Shipping Containers." (See Reference 21.)

at 2 Hz and allowed to sweep up to 100 Hz or 150 Hz at a constant input g level (0.5 g's is usually sufficient). The sweep up and back is allowed to occur twice and the natural frequencies of the specimen noted. The vibration table is then set to dwell at each of the natural frequencies of the specimen for fifteen minutes or more. The product and package are inspected for damage following each dwell.

Full sized loads used in shipping are used in the **Unitized Load** or **Vertical Shock Resonance Test.** Alternately, a single vertical column equal to the height used in actual shipment may be used as the test specimen. The specimen is restrained using fixtures to allow some horizontal movement and free vertical movement. The vibration table is used to sweep up and down between 2 Hz and 100-150 Hz twice and then dwell at the resonant frequencies of the stack. Products and packages are then inspected for damage. The test should reveal whether the container has sufficient strength to survive dynamic compression loads.

The interpretation of the results of these tests is an art, developed by the packaging technologist with time and experience.

5.3 Random Vibration

In reality the vibrational forces experienced by products and packages in distribution are not the result of single frequency, sinusoidal vibration. The vibration of a truck bed, for example, is random in nature as shown in the graph of Figure 5.6. This random vibration may be thought of as a mixture or sum of several sinusoidal vibrations of different frequencies and amplitudes. The product may be excited by several sinusoidal vibrations at any single instant, which may have a completely different effect from the single frequency excitations we have explored up to now. Testing the product package system with single frequency excitation may not always be adequate to uncover all of the possibilities for damage.

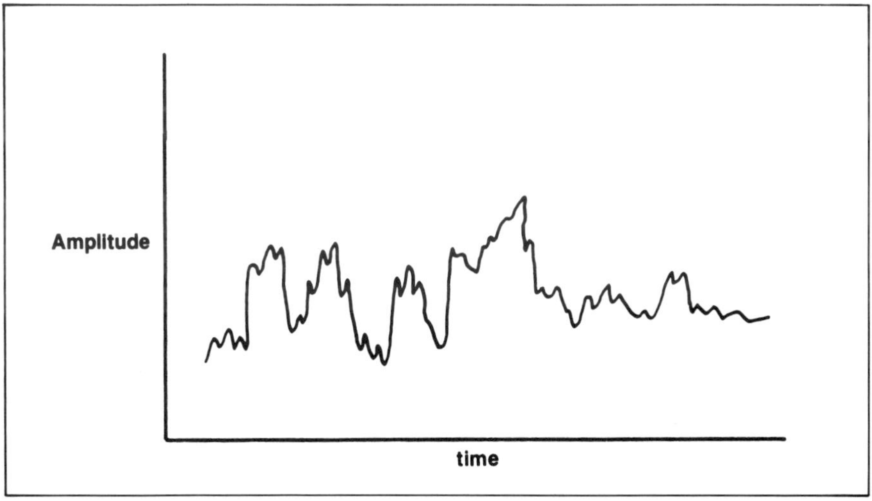

Figure 5.6. Random Vibration of a Truck Bed.

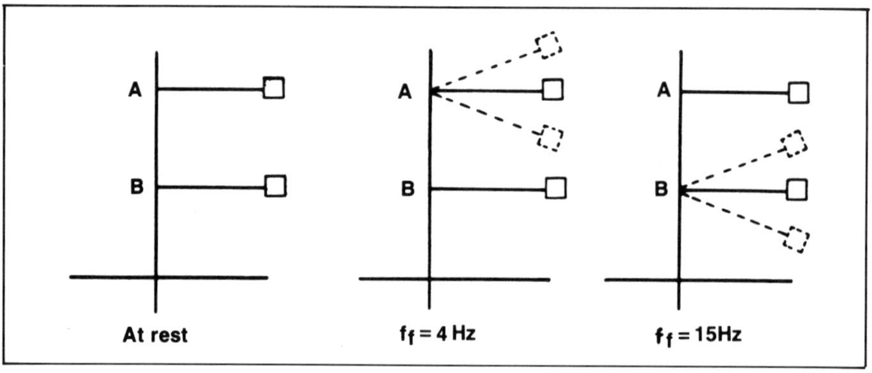

Figure 5.7. Response of Elements to Single Forcing Frequencies.

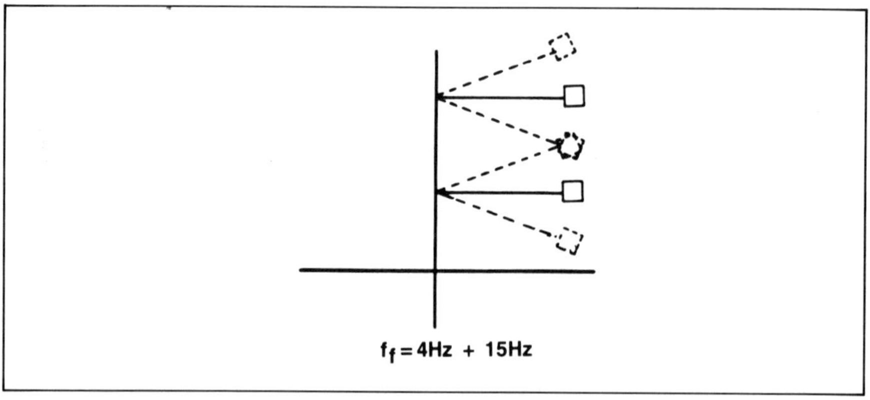

Figure 5.8. Response of Elements to Multiple Forcing Frequencies.

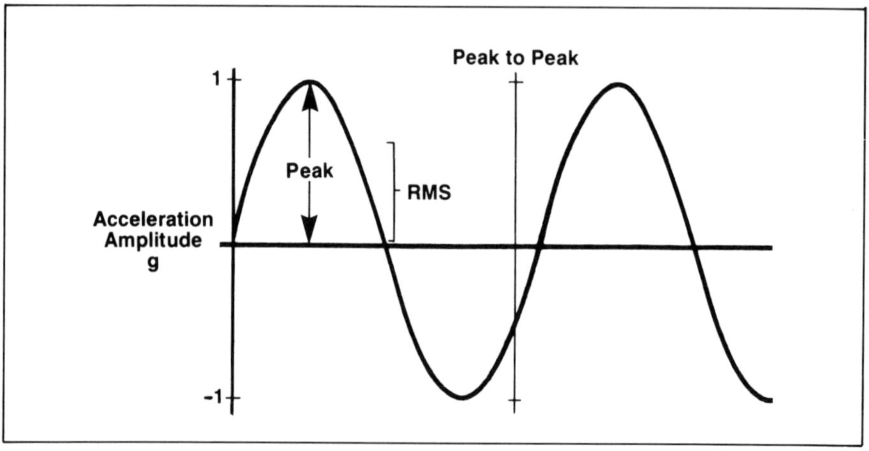

Figure 5.9. Sinusoidal Vibration.

Consider a product with two critical elements A and B. Element A has a natural frequency of 4 Hz and Element B, a natural frequency of 15 Hz. Figure 5.7 illustrates the response of this product to either 4 Hz or 15 Hz sinusoidal forcing frequencies. Elements A and B are spaced far enough apart so that when only one of them is vibrating they do not touch and will not damage one another. However, if the product is subjected to 4 Hz and 15 Hz vibrations simultaneously, as shown in Figure 5.8, the elements may hit one another, causing damage. This type of damage would not be uncovered using the single frequency vibration tests previously discussed. Many packaging technologists believe that product-package systems should be subjected to random vibration tests to reveal such possibilities.

5.4 Fourier Analysis

Let's consider the sinusoidal vibration shown in Figure 5.9. The frequency, as we have seen before, is the reciprocal of the period measured in seconds. There are three different ways of expressing the acceleration here:

Peak Acceleration Amplitude $= 1$ g

Peak to Peak Acceleration Ampliltude $= 2$ g

Root Mean Square (RMS) Acceleration Amplitude $= \frac{1}{\sqrt{2}}$ x peak amplitude

$= 0.707$ x peak amplitude

$= 0.707$ g

The RMS can be thought of as the effective energy of the vibration.

Another method of representing vibration, the frequency domain representative, is shown in Figure 5.10. This is simply the RMS amplitude plotted against the vibration frequency.

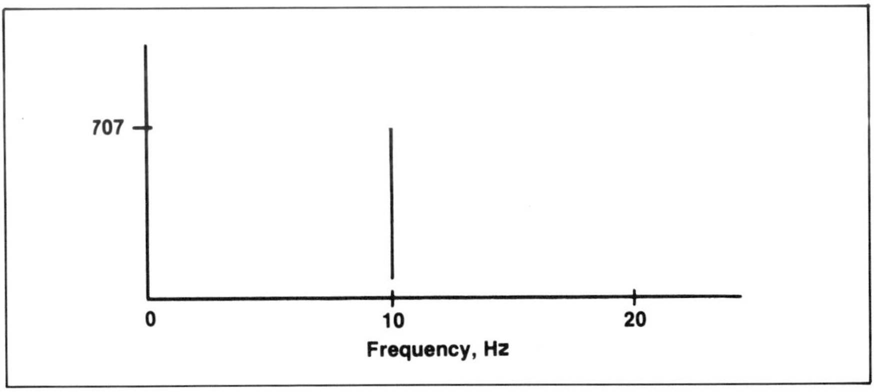

Figure 5.10. Frequency Domain Representation of Vibration.

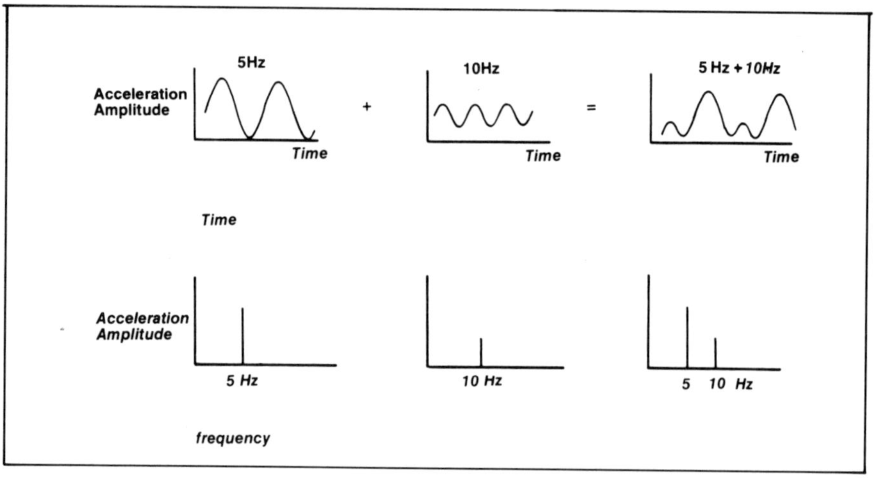

Figure 5.11. The Addition of Sine Waves.

Baron Fourier demonstrated over one hundred years ago that any complex waveform may be generated by adding sine waves with the appropriate frequencies and amplitudes. Figure 5.11 illustrates such a sum for both the time and frequency domains. More useful is the fact that this process may be applied in reverse to break a complex waveform down into simple sine wave components. This process is shown in Figure 5.12, and is commonly referred to as Fourier Analysis.

There are many approaches to Fourier analysis, among which the use of the Band Pass Filter (BPF) is the most straightforward. The BPF is an electronic device which breaks a complex waveform down into its sine wave components, referred to as the Fourier transformation. It is the nature of the BPF to allow only a very narrow band of frequency to pass and block all other frequencies, as shown in Figure 5.13. The bandwidth of the BPF is the difference between the highest and lowest frequencies that can pass through the filter. For example, the first BPF in Figure 5.13 has upper and lower frequency ranges of 5.5 Hz and 4.5 Hz, respectively, giving it a bandwidth of 1 Hz. If we use enough BPF's to cover a broad frequency range (i.e. 200 Hz for truck vibration), any seemingly complex waveform may be transformed into a comprehensible set of sine waves with known frequencies and amplitudes.

5.5 The Power Density Spectrum

We can decompose the complex waveform representing the random vibration of a truck bed into a Fourier representation of simple sine waves using Band Pass Filters of appropriate construction. This representation is a close approximation of the real waveform with one important shortcoming. The peak amplitude of any component sine frequency in the Fourier representation is a constant, while one of the characteristics of the random vibration we are trying to represent is that the amplitude associated with any frequency is a constantly varying quantity. For example, the peak acceleration amplitude associated with the 5 Hz component may be 0.5 g one moment and 0.7 g a moment later. To incorporate this random variation into the representation of the complex motion we must turn to statistics.

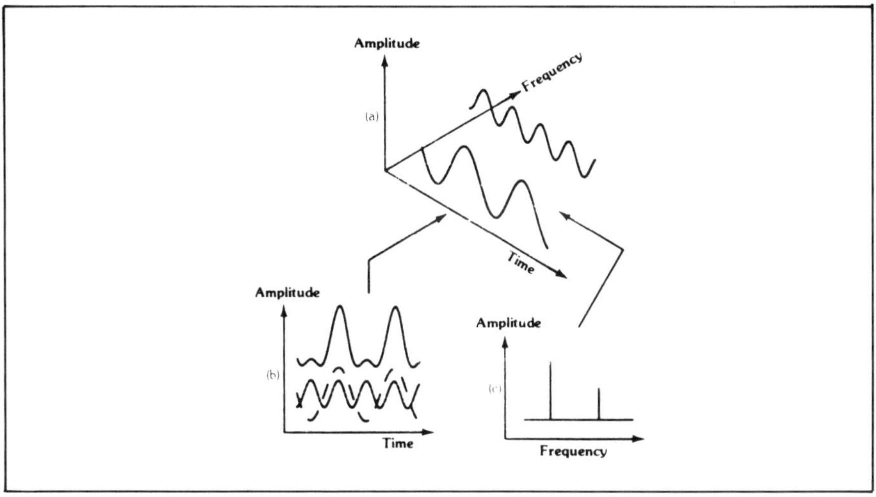

Figure 5.12. Fourier Analysis. (20)

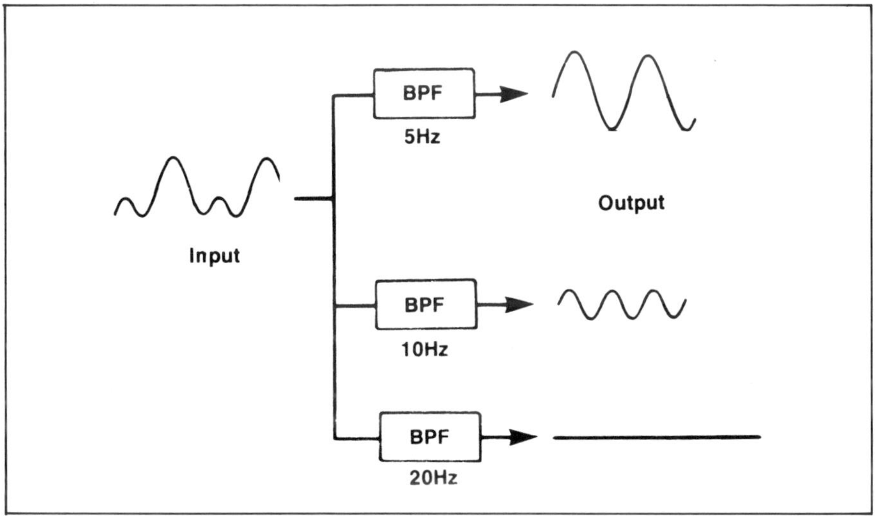

Figure 5.13. Fourier Analysis Using Band Pass Filters (BPFs). (20)

The Normal or Gaussian statistical distribution may be employed to describe the instantaneous peak acceleration amplitude of any of the component frequencies composing the random vibration. The amplitude of any of the component frequencies may be represented as the **Power Density:**

$$\text{P.D.} = \frac{\sum_i (\text{RMS})_i^2/N}{\text{BW}} \quad \frac{g^2}{\text{Hz}} \qquad \textbf{5.5}$$

where:

> **RMS_i** is the Root Mean Square acceleration value measured in g's at any instant, i;
>
> **N** is the number of instants sampled, and;
>
> **BW** is the bandwidth of the BPF used to obtain the component frequency (usually normalized to be 1 Hz).

A typical plot of Power Density versus frequency, also known as the **Power Density Spectrum,** for a truck bed is shown in Figure 5.14.

In statistical terms, the power density at any given frequency is the variance about a mean value of zero g. Therefore, based on the probabilities associated with the normal distribution, we may predict the acceleration levels associated with any of the component frequencies of the complex waveform:[9]

±1 PD values occur 68.3% of the time,

±2 PD values occur 95.4% of the time,

±3 PD values occur 99.7% of the time.

The interpretation of the Power Density Spectrum may be illustrated with a simple example:

From Figure 5.14, the Power Density at 5 Hz is $10^{-2}\ g^2/Hz$. The bandwidth has been normalized to 1 Hz, so it may be ignored.

$$PD = \sqrt{10^{-2}\ g^2} = 0.10\ g \text{ (the RMS value)}$$

since

$$\text{RMS Acceleration Amplitude} = 0.707 \text{ Peak Amplitude,}$$

$$\text{Peak Acceleration Amplitude} = \frac{0.10g}{0.707} = 0.14\ g$$

For the truck ride represented in Figure 5.14, we may expect peak acceleration levels between:

+0.14 g and -0.14 g 68.3% of the ride,

+0.28 g and -0.28 g 95.4% of the ride, and

+0.42 g and -0.42 g 99.7% of the ride.

This is fairly consistent with the common wisdom that the worst vertical acceleration of a truck ride is 0.5 g at low frequencies.

Random vibration and PSD are discussed further in References 1, 9, and 13.

---

[9]Based on the probabilities of occurrence of the standard deviation, where:

$$\text{Standard Deviation} = \sqrt{\text{Variance}} = \sqrt{\text{Power Density}}$$

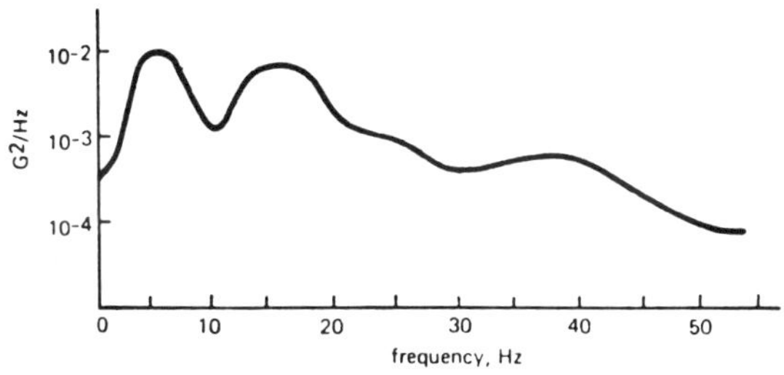

Figure 5.14. A Power Density Spectrum. (3)

5.6 Random Vibration Test Equipment

The random vibration encountered in distribution is literally not random because the acceleration levels associated with each component frequency is statistically predictable, and the presence or absence of a given frequency is a characteristic of the mode of distribution and the vehicles involved. The simulation of the complex vibrations found in distribution is not such a simple task.

Hydraulic or electromechanical vibration tables are capable of producing either single frequency sinusoidal vibration or random vibration. However, the control equipment for random vibration is considerably more complex and expensive than for simple vibration.

Let's consider the control mechanism for sinusoidal vibration shown in Figure 5.15. Here, a single frequency with a given amplitude is applied to the table. An accelerometer mounted on the table monitors the table's vibration. If the table's vibration is different from the desired vibration, as judged by the comparator, a correction is made on the input signal to the table. The control mechanism and comparator are needed to make the table vibrate at the desired frequency and amplitude with the desired wave shape because the table's response to the input signal is always far from perfect. The constant feedback and correction process keeps the vibration as close to that desired as is mechanically possible. However, at any instant during the table's vibration, the control mechanism works on only one frequency.

In random vibration the control mechanism must work with a spectrum of frequencies at any one instant. Currently, there are three schemes for producing random vibration in the laboratory:

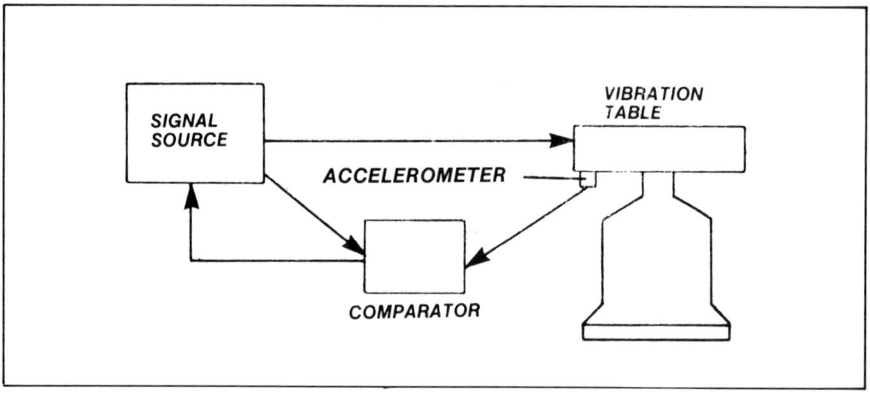

Figure 5.15. A Control Mechanism for Producing Single Frequency Sinusoidal Vibration. (9)

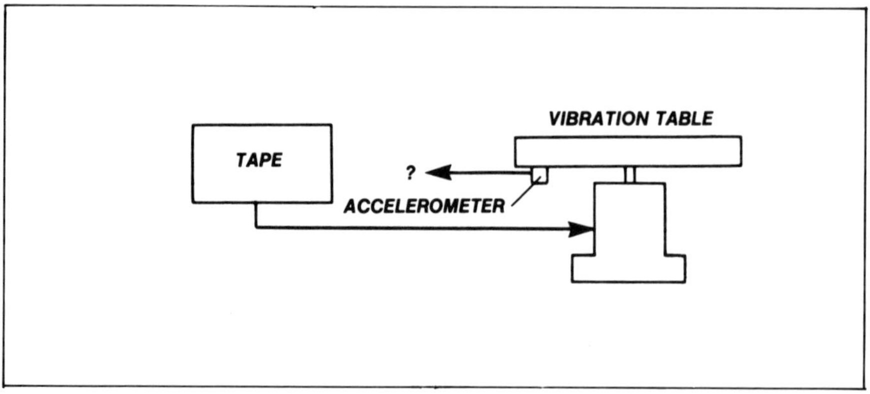

Figure 5.16. Tape Recorded Random Vibration Control. (9)

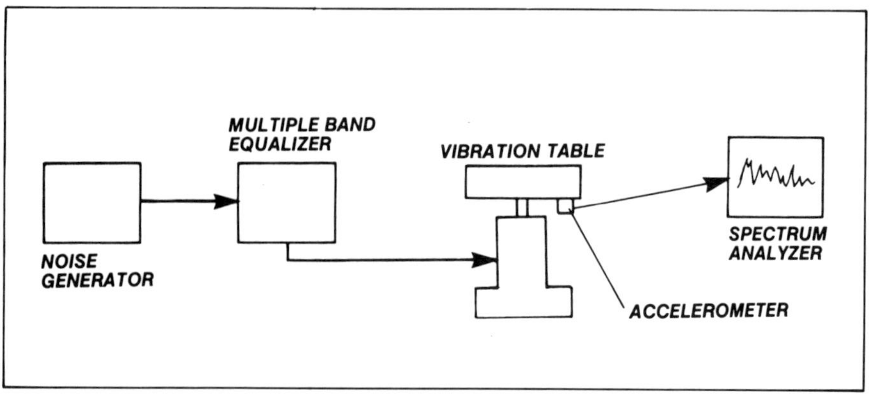

Figure 5.17. A Manually Controlled Analog System for Random Vibration. (9)

1. **Tape Recorded Random Vibration Control.** As shown in Figure 5.16, a prerecorded signal is fed into the vibration control system producing a response vibration by the table. Since the response of the table to an input signal is not perfect, the recorded signal must be made specifically for each table with a particular load.[10]

2. **Manually Controlled Analog System for Random Vibration Control.** This system is shown in Figure 5.17. Here a noise generator produces a truly random signal. The multiple band equalizer selects the desired frequencies and amplitudes from the generator's signal and shapes the signal input to the control mechanism for the vibration table. The vibration of the table is monitored visually using the spectrum analyzer which displays the Power Density Spectrum of the vibrating table. The multiple band equalizer is manually adjusted to obtain the desired PSD output on the spectrum analyzer.

3. **Automatic Digital Control.** Shown in Figure 5.18, the equipment used in this method is different from that used in the manually controlled system above, but the approach is the same. Here, a digital computer generates an input signal, monitors the output signal from the vibrating table and automatically corrects the input so that the table's vibrations conform to some preselected PSD.

The applications of random vibration to package design and testing is still in its infancy. We expect considerable advances in this area in the next few years which should add a greater sophistication to the tools available to the packaging technologist.

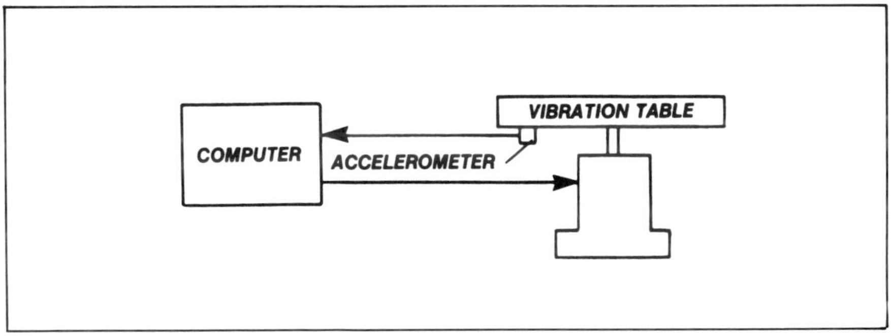

Figure 5.18. Automatic Digital Control for Producing Random Vibration. (9)

[10]Because the load changes the mass of the table-load system, the response of the system varies with the load.

6. MECHANICAL SHOCK

6.0 Introduction

Throughout the distribution system packages are manhandled and mishandled in various ways. They are dropped, thrown, kicked and otherwise roughly abused. Packages may fall from conveyors or forklifts and crash to the floor. They are also subjected to a variety of vehicle impacts; trucks starting, stopping, hitting chuckholes and railroad crossings, railcar humping, jolting and other moderately violent actions. In each instance the package suffers an impact with another object; floor, truck bed, pallet, bulkhead or another package. This impact results in a **mechanical shock** to both objects.

A mechanical shock occurs when an object's position, velocity or acceleration suddenly changes. Such a shock may be characterized by a rapid increase in acceleration followed by a rapid decrease over a very short period of time. The acceleration versus time plot for most shocks is very complex, as shown in Figure 6.1a. Shocks are more easily understood if we replace the complex history of Figure 6.1a with the smooth approximation shown inFigure 6.1b. A package shock may typically be 20 milliseconds (0.020 seconds) long and have a magnitude or "height" of 150 g's.

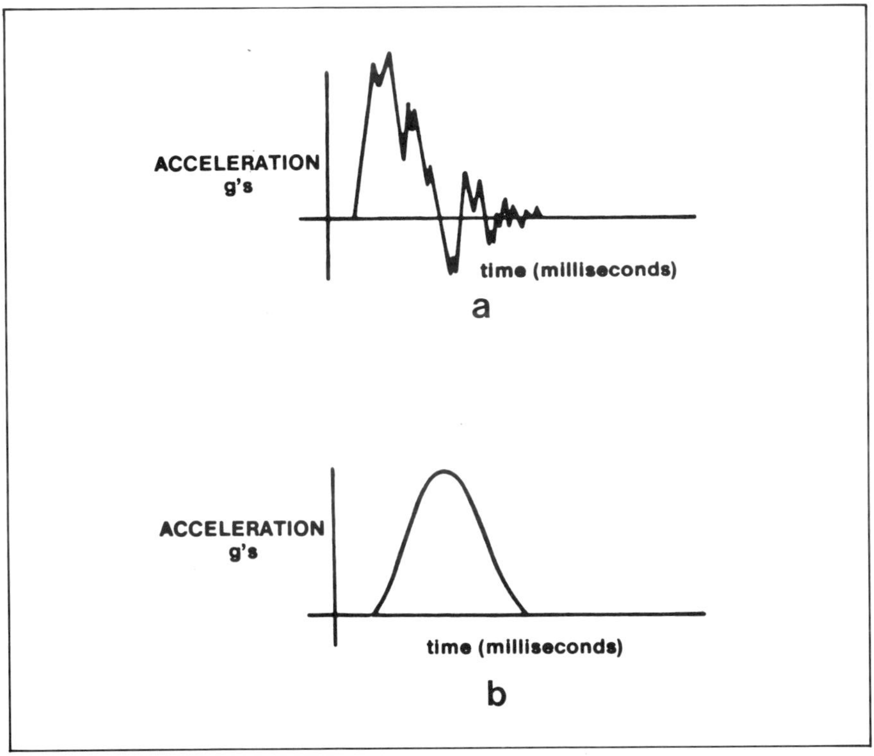

Figure 6.1. Representation of Mechanical Shock.

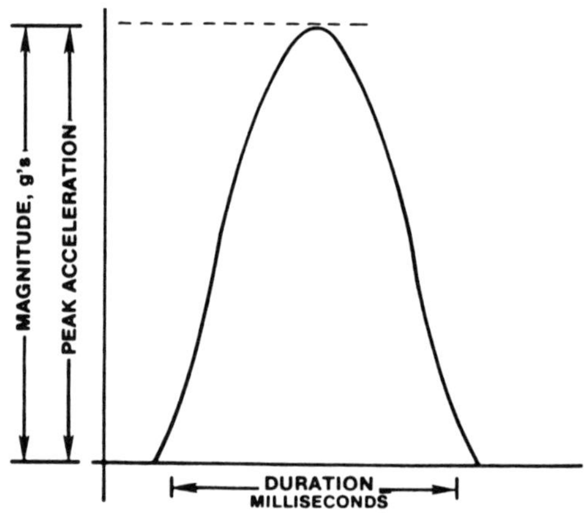

Figure 6.2. Parameters Describing a Mechanical Shock.

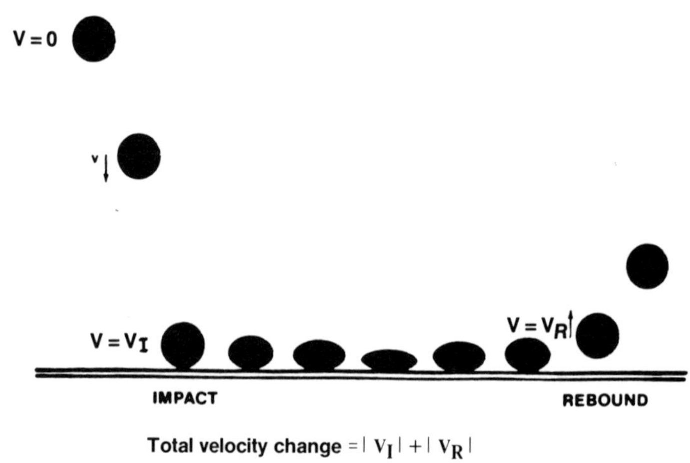

Figure 6.3. A Falling Package.

A mechanical shock may be thought of as a multiple of weight applied suddenly and lasting only a fraction of a second. If the author were to receive a 200 g shock, for instance, he would feel as if he weighed 45,000 pounds for an instant. This would feel like nothing more than a minor jolt due to the natural shock absorbing muscle and tendon system of the body (if, however, the shock were to last longer, the consequences would be considerably worse). Mechanical shocks to packages, however brief, may cause damage.

To understand and estimate the potential damage a shock may cause, we need to know both the magnitude of the acceleration and the duration of the shock.

6.1 The Free Falling Package

In Section 2.4, the motion of a freely falling body was examined. Equations 2.6 and 2.7, respectively describe the length of time it takes a package to fall from a drop height, h, and the downward velocity at which it will be traveling a moment before impact; v_I, the **impact velocity:**

$$t = \sqrt{\frac{2h}{g}} \qquad 2.6$$

$$v_I = \sqrt{2\,gh} \qquad 2.7$$

Such a fall is shown in Figure 6.3. From experience, we know that a small object like a package impacting a large object like a floor will bounce. The amount of bounce depends on the nature of the objects involved. A bean bag dropped on a concrete floor will not rebound, whereas a "superball" will rebound nearly to the point from where it was dropped. Likewise, a package will rebound a little or a lot depending on the nature of the package and the surface it hits.

The **coefficient of restitution**[1], e, describes the **rebound velocity** as a function of the impact velocity:

$$v_R = e\,v_I \qquad 6.1$$

The coefficient e takes on values between zero and one, with typical values falling in the 0.3 to 0.5 range. Note that if e were to be greater than one, the rebound velocity would exceed the impact velocity and the package would bounce higher than the drop point.

The total **velocity change, Δv**, for an impact is defined as the sum of the absolute values of the impact and rebound velocities:

$$\Delta v = |v_I| + |v_R| \qquad 6.2$$

or, since we know e and v_I:

$$\Delta v = (1 + e)\,v_I = (1 + e)\sqrt{2\,gh} \qquad 6.3$$

---

[1]For a lightly damped system, $e \cong 1 - \pi c/(2\sqrt{km})$, where c is the damping coefficient.

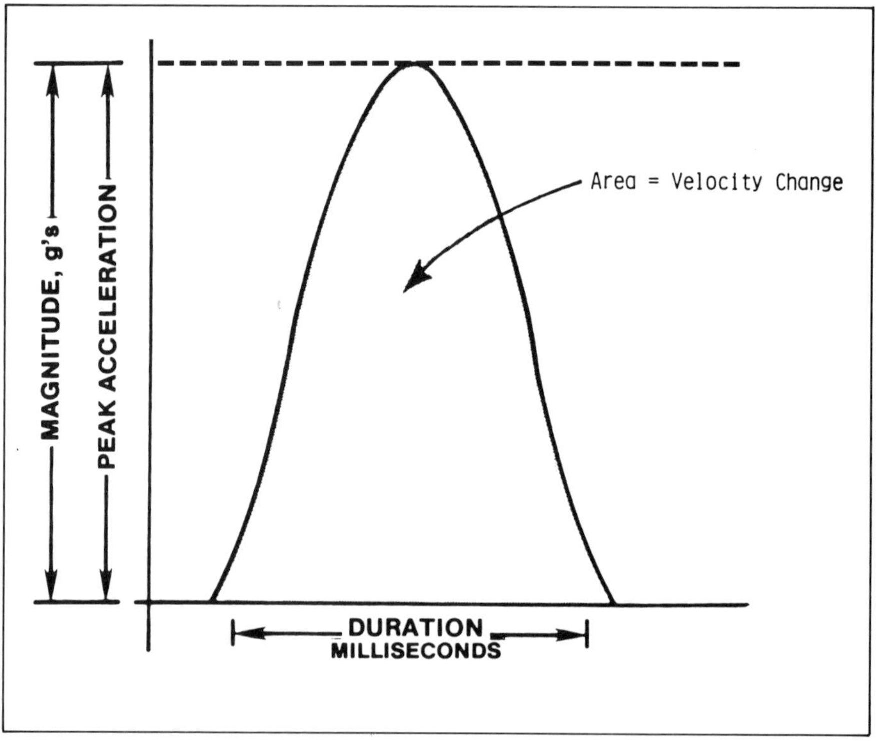

Figure 6.4. The Relationship Among Shock Parameters.

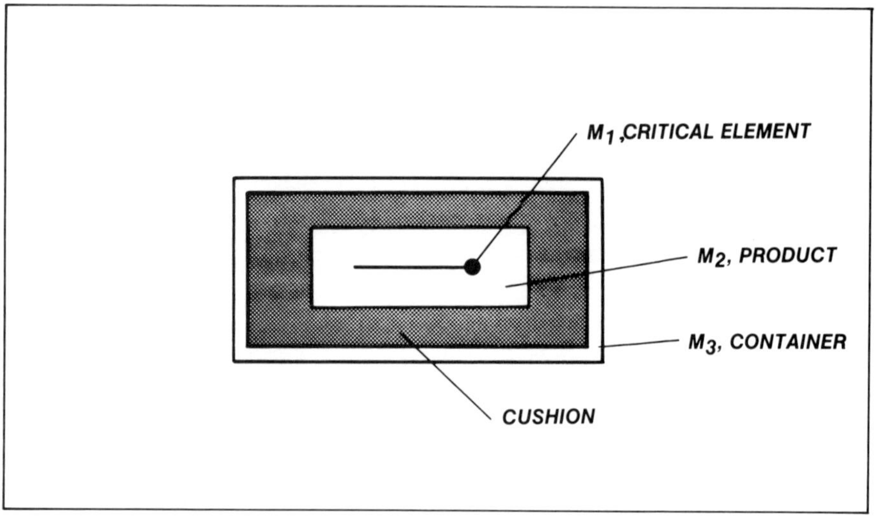

Figure 6.5. A Simple Product-Package System. (10)

Because 0 $\leqslant e \leqslant 1$;

$$\sqrt{2gh} \leqslant \Delta v \leqslant 2\sqrt{2gh} \qquad 6.4$$

From Equation 6.4 we may estimate the velocity change resulting from an impact to be somewhere between the value of the impact velocity and twice that value.

The velocity change is also numerically equal to the area beneath the shock pulse as shown in Figure 6.4.

Package damage is related to the three factors involved in mechanical shock:

Peak Acceleration
Duration
Velocity Change

Knowing any two of the factors allows us to estimate the third.

6.2 Mechanical Shock Theory

Mindlin (10), in considering distribution breakage of radar tubes during World War II, reduced the product-package system to its simplest form. Shown in Figure 6.5, the product-package system consists of four basic components: the outer container, the cushion, the product, and a critical element. The critical element is the most fragile component of the product (the filament of the radar tube, for instance). It is the part that is most easily damaged by mechanical shock.

This model may be further reduced to something we may investigate mathematically, shown in Figure 6.6. This model and the following analysis will help in understanding the potentials for product shock damage. For our product-package model:

M_2 represents the mass of the product

M_1 represents the mass of the critical element or CE

M_3 represents the mass of the outer container

k_1 is the linear spring constant of the spring-mass system representing the critical element

k_2 is the linear spring constant of the cushion system[2]

For simplicity we will ignore the mass of the outer container and assume that it provides no spring action. We will also assume that the cushion has no mass or damping and suffers no permanent deflection from a shock. Also we will assume that the product-package system impacts a perfectly rigid floor and that:

$$M_1 \ll M_2$$

[2]We will use a simple linear spring to represent the cushion to simplify our analysis. Other nonlinear models may be more appropriate, but also entail complex mathematics.

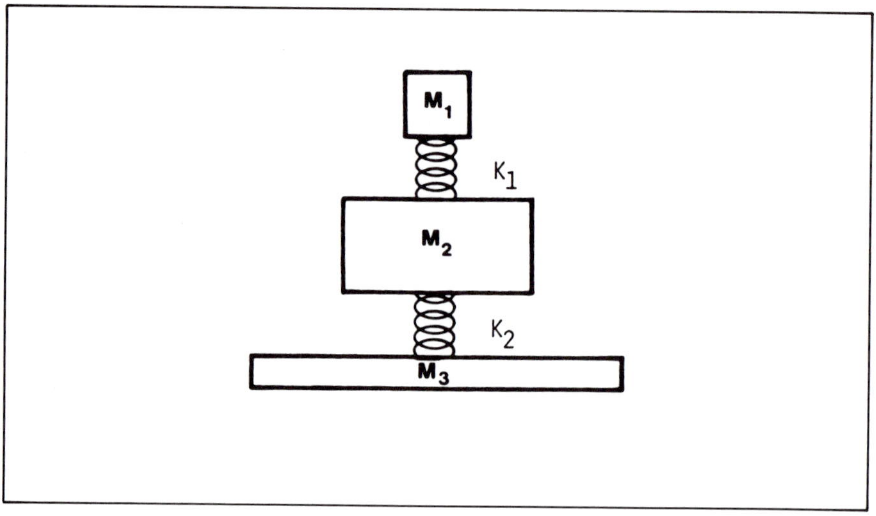

Figure 6.6. A Spring-Mass Model for the Product-Package System Shown in Figure 6.5. (10)

That is, the mass of the critical element is negligible compared to the mass of the product.

In Figure 6.7, the product-package is shown raised to a height, h, and poised to drop. The critical element may be ignored for the moment as we investigate the shock to the product.

At Point A, the product is ready to fall. The potential energy at this moment is:

$$PE = M_2gh = W_2H \qquad 6.5$$

where the weight of the product is given by W_2 ($W_2 = M_2g$).

The product-package system is allowed to fall. At Point C, the outer container makes contact with the impact surface. At this point, before the product begins to compress the cushion, we set up a reference line so that we may measure the product's downward displacement, x_2, on the cushion. Also at Point C, the kinetic energy of the system is given by

$$KE = 1/2\, M_2\, v_I^2 \qquad 6.6$$

where the impact velocity is given by Equation 2.7.

Substituting

$$KE = 1/2\, M_2(2\, gh)^2 = M_2gh = W_2h \qquad 6.7$$

we see that the kinetic energy at this moment is equal to the initial potential energy of the system.

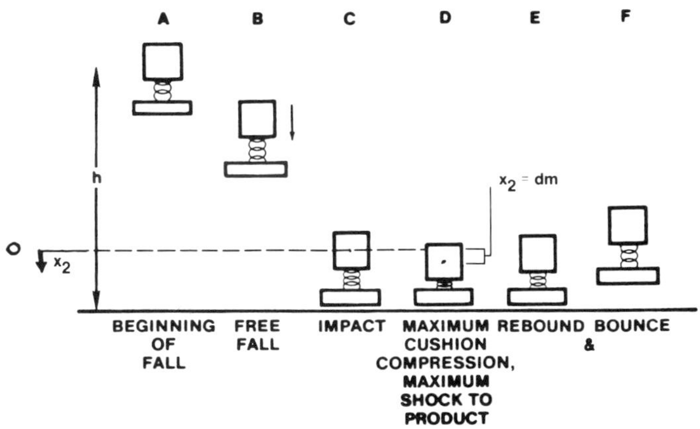

Figure 6.7. The Impact of a Product-Package System.

In the moments following the initial contact of the container with the impact surface the cushion must absorb this kinetic energy. In this situation the amount of energy absorbed by the linear cushion is,

$$E = 1/2\, k_2 x_2^2 \,, \qquad 6.8$$

where x_2 is the downward displacement of the product on the cushion.

A Point D the cushion has absorbed all of the system's kinetic energy and the downward velocity of the product has slowed to zero. The **maximum deflection, d_m** of the cushion occurs at this point:

$$\text{maximum dynamic compression, } d_m = \text{maximum value of } x_2$$

Equation 6.7 then becomes

$$E_{max} = 1/2\, k_2\, d_m^2 \qquad 6.9$$

Since the maximum amount of energy absorbed by the cushion must equal the system's kinetic energy at impact, we may equate expressions 6.7 and 6.9:

$$KE = W_2 h = E_{max} = 1/2\, k_2\, d_m^2$$

$$W_2 h = 1/2\, k_2\, d_m^2 \qquad 6.10$$

Equation 6.10 may be rearranged to produce an expression for the maximum dynamic compression:

$$d_m = \sqrt{\frac{2\, W_2 h}{k_2}} \qquad 6.11$$

Recall from Equation 3.11 that the **static** deflection of the product on the linear cushion is:

$$\delta_{st} = \frac{W_2}{k_2}$$

so Equation 6.11 may also be written as:

$$d_m = \sqrt{2h\,\delta_{st}} \qquad 6.12$$

The maximum force P_{max} exerted upward by the cushion against the product occurs when $x_2 = d_m$:

$$P_{max} = k_2 x_2 = k_2 d_m = k_2\sqrt{\frac{2\,W_2 h}{k_2}}\,,$$

$$P_{max} = \sqrt{2\,k_2\,w_2 h} \qquad 6.13$$

The maximum acceleration (or deceleration) experienced by the product, G_m, may be found from the relationship[3]:

$$G_m = \frac{P_{max}}{W_2} \qquad 6.14$$

$$G_m = \sqrt{\frac{2\,k_2 W_2 h}{W_2}}\,,$$

$$G_m = \sqrt{\frac{2\,k_2 h}{W_2}} \qquad 6.15$$

where G_m is unitless, but is understood to be in units of 1 g.

Again, recalling the expression for δ_{st}, an alternate form of Equation 6.15 may be written:

$$G_m = \sqrt{\frac{2h}{\delta_{st}}} \qquad 6.16$$

---

[3]At the point of maximum compression there is no motion, the upward and downward forces are in equilibrium:

$$M_2\ddot{x}_2 + P_{max} = 0,$$

$$M_2\ddot{x}_2 = -P_{max}$$

$$\frac{W_2}{g}\ddot{x}_2 = -P_{max}$$

$$W_2\frac{\ddot{x}_2}{g} = -P_{max}$$

$$W_2 \left/\frac{\ddot{x}_2}{g}\right/ = P_{max}$$

$$W_2\,G_m = P_{max} \text{ where } G_m = \left/\frac{\ddot{x}_2}{g}\right/$$

$$G_m = \frac{P_{max}}{W_2}$$

Notice that in Equations 6.15 and 6.16, G_m is proportional to the square root of the drop height:

$$G_m \propto \sqrt{h}$$

This means that if we double the drop height, the magnitude of the shock is **not** doubled, but rather increased by a factor of $\sqrt{2}$ (about 1.4). Equation 6.15 may be rearranged to produce an expression for k_2:

$$k_2 = \frac{W_2 G_m^2}{2h} \quad 6.17$$

Equation 6.11 may also be rearranged to produce an equivalent expression:

$$k_2 = \frac{2 W_2 h}{d_m^2} \quad 6.18$$

Equations 6.1 and 6.15 may also be manipulated to give us a simple expression for d_m. Multiplying both sides of Equation 6.11 by 1:

$$(1)\, d_m = \sqrt{\frac{2W_2h}{k_2}} \left(\sqrt{\frac{2h}{2h}}\right),$$

$$d_m = \sqrt{\frac{2\, W_2h\, 2h}{k_2\, 2h}},$$

$$d_m = 2h\sqrt{\frac{W_2}{2\, k_2h}},$$

$$d_m = \frac{2h}{\sqrt{\frac{2k_2h}{W_2}}},$$

$$d_m = \frac{2h}{G_m}, \quad 6.19$$

giving us an expression for the maximum dynamic compression as a function of the maximum acceleration and the drop height. Several examples will illustrate how these equations may be used.

1. Find d_m and k_2 for a cushion which will limit the maximum acceleration of an 82 lb. product to 62 g's in an eleven foot drop.

   Here we are given that:

   W_2 = 82 lb.

   h = 11 ft. = 132 in.

   G_m = 62

From Equation 6.19:

$$d_m = \frac{2h}{G_m} = \frac{2 \cdot 132 \text{ in.}}{62} = 4.26 \text{ in.}$$

From Equation 6.17:

$$k_2 = \frac{W_2 G_m^2}{2h} = \frac{82 \text{ lb. } (62)^2}{2 \cdot 132 \text{ in.}} = 1194 \text{ lb/in.}$$

We may conclude that an 82 lb. product cushioned so that k_2 is equal to 1194 lb/in. and falling eleven feet will experience a shock of 62 g's and depress the cushion 4.26 inches. Obviously, enough cushion must be present to absorb a deflection equal to or greater than d_m.

2. A one hundred pound product measures 10 in. x 10 in. x 10 in. and can sustain up to a 50 g shock without damage. It is equally sensitive to shock on all six faces. If a maximum drop height of five feet is expected in distribution, what is the modulus of elasticity of the required cushion material? Assume that the working length of the cushion material is fifty percent of the cushion's total thickness.

We are given that:

$$W_2 = 100 \text{ lb.}$$

$$h = 5 \text{ ft.} = 60 \text{ in.}$$

$$G_m = 50$$

First, we need to calculate d_m:

$$d_m = \frac{2h}{G_m} = \frac{2.60 \text{ in.}}{50} = 2.4 \text{ in.}$$

We are to assume that for the cushion:

Working length = 50% total thickness

$$WL = 0.5 \text{ TT}$$

In order to prevent "bottoming out:"

$$WL \geqslant d_m$$

Setting the maximum dynamic compression equal to the working length:

$$d_m = WL = 0.5 \text{ TT}$$

$$TT = \frac{d_m}{0.5} = \frac{2.4}{0.5} = 4.8 \text{ in.}$$

The cushion must be 4.8 inches thick.

From Equation 3.26, the modulus of elasticity, E_2, is related to the spring constant:

$$k_2 = E_2 \left(\frac{\text{Area}}{\text{Thickness}}\right)$$

And k_2 for our problem is:

$$k_2 = \frac{W_2 G_m^2}{2h} = \frac{100 \text{ lb. } (50)^2}{2 \cdot 60 \text{ in.}} = 2083 \text{ lb/in.}$$

So:

$$2083 \text{ lb./in.} = E_2 \left(\frac{\text{Area}}{\text{Thickness}}\right)$$

The thickness of each cushion is 4.8 inches and the bearing area is equal to the area of each face of the product, 100 in^2.

$$2083 \text{ lb./in.} = E_2 \left(\frac{100 \text{ in}^2}{4.8 \text{ in.}}\right)$$

$$E_2 = (2083 \text{ lb./in.}) \left(\frac{4.8 \text{ in.}}{100 \text{ in}^2}\right)$$

$$= 100 \text{ lb./in}^2$$

We also may want to know the size of the container that this ten inch cubic product and its cushions need for distribution. If we place a 4.8 inch thick cushion on each of the product's six faces, each dimension will be increased by the thickness of two cushions:

10 in. + 4.8 in. + 4.8 in. = 19.6 inches

The inner dimensions of the container are then:

19.6 in. x 19.6 in. x 19.6 in.

Before leaving this section, there is one more interesting relationship to cover. We expect a maximum acceleration of G_m to produce a displacement of d_m. Under normal conditions the one g acceleration of gravity produces a displacement of δ_{st}. Therefore, we would expect the relationship below to hold:

$$\delta_{st} = \frac{d_m}{G_m}$$

Therefore by substituting:

$$\frac{d_m}{G_M} = \frac{\sqrt{\dfrac{2W_2h}{k_2}}}{\sqrt{\dfrac{2k_2h}{W_2}}} = \sqrt{\frac{2\,W_2h}{k_2} \cdot \frac{W_2}{2k_2h}}$$

$$= \sqrt{\frac{W_2^2}{k_2^2}} = \frac{W_2}{k_2} = \delta_{st}\,,$$

we see that it does.

In this section we have explored the relationships of G_m, d_m, and k_2. Even though our model is based on a theoretical linear spring we now know:

1. G_m is related to the drop height, the springiness of the cushion and the weight of the product.

2. The stiffer the spring or cushion, the larger the value for G_m.

3. The higher the drop height, the larger the value for G_m ($G_m \propto \sqrt{h}$).

4. The heavier the product, the smaller the value of G_m, as long as the spring or cushion is working.

6.3 Shock Duration

Packages and products typically receive mechanical shocks lasting somewhere between one and fifty milliseconds in the distribution system. The time length or duration of the shock is referred to as τ [4].

We may approximate the displacement of the product on the cushion, x_2, at any time during the impact with the function:

$$x_2(t) \simeq d_m \sin (W_2 t) \qquad 6.20$$

shown in Figure 6.8, where $W_2 = 2\pi f_2$, and

$$f_2 = \frac{1}{2\pi}\sqrt{\frac{k_2 g}{W_2}} \qquad 6.21$$

the natural frequency of the product (M_2) on the cushion (k_2). The period of the free vibration[5] of the product on the cushion, T_2, may be seen from Figure 6.9 to be twice the shock duration:

$$\frac{1}{f_2} = T_2 = 2\tau \qquad 6.22$$

[4] TAU

[5] Think of the shock pulse as one half of a vibration cycle.

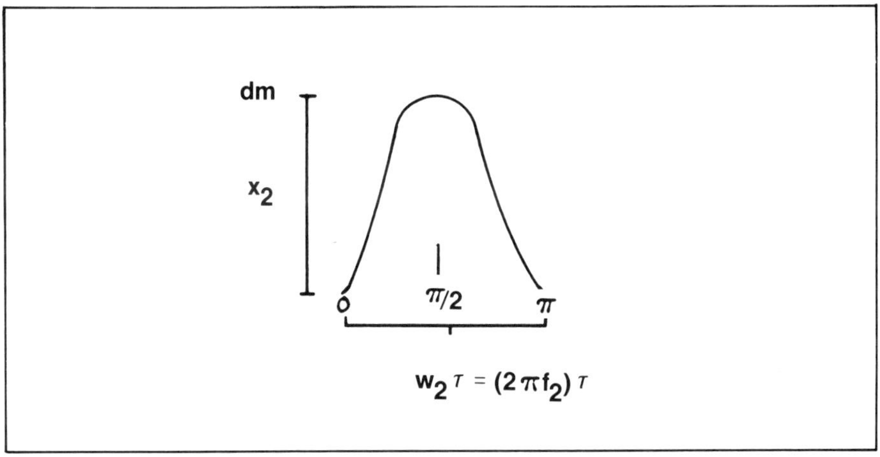

Figure 6.8. Shock Duration and Maximum Displacement.

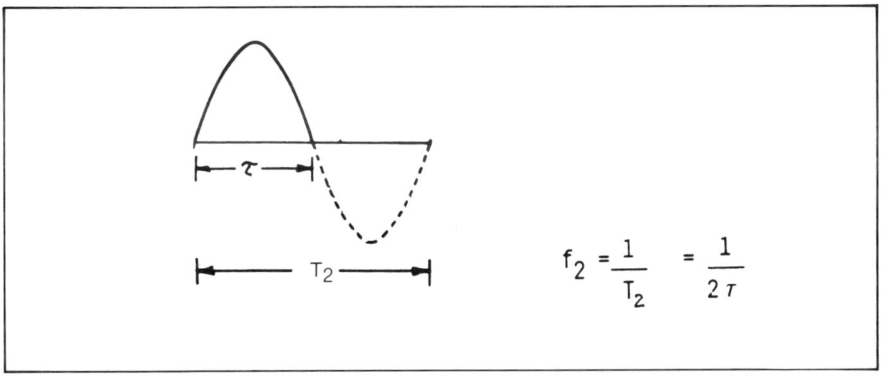

Figure 6.9. Shock Duration and the Natural Period of Vibration.

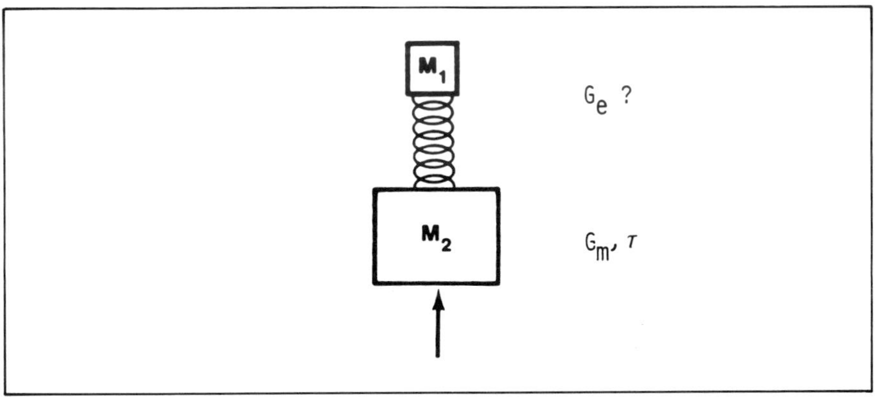

Figure 6.10. Shock Transmission.

For any shock of known period τ, we may calculate the **equivalent shock frequency**[6];

$$f_2 = \frac{1}{2\tau} \qquad 6.23$$

Equations 6.21 and 6.23 may be rearranged to give us an expression for the shock duration:

$$\tau = \frac{1}{2f_2} = \frac{1}{2\left(\frac{1}{2\pi}\sqrt{\frac{k_2 g}{W_2}}\right)} = \frac{\pi}{\sqrt{\frac{k_2 g}{W_2}}} \qquad 6.24$$

$$\tau = \pi\sqrt{\frac{W_2}{k_2 g}} \qquad 6.25$$

In example 1 of Section 6.2 we had an 82 lb. product and a calculated spring constant of 1194 lb./in. The shock duration is then:

$$\tau = \pi\sqrt{\frac{82 \text{ lb.}}{(1194 \text{ lb/in})(386.4 \text{ in/sec}^2)}} = 0.042 \text{ seconds}$$

or 42 milliseconds. Notice that based on the approximations of Equation 6.20, the shock duration is **independent** of drop height,

6.4 Shock Amplification and the Critical Element

To adequately understand the shock phenomenon and ultimately protect a product from shocks encountered in distribution, we must compare the response of the critical element within the product to the shock occurring to the product during impact. In the last two sections, expressions for G_m and τ were developed. Now we look at what happens to the critical element as a result. We will define G_e as the maximum acceleration experienced by the critical element, resulting from a shock to the product of magnitude G_m and duration, τ, as shown in Figure 6.10.

Analogous to our approach to vibration magnification we may define an **amplification factor,** A_m, relating the input and output shock levels:

$$A_m = \frac{G_e}{G_m} \qquad 6.26$$

The maximum acceleration at the critical element is simply the product of this amplification factor and G_m:

$$G_e = A_m G_m \qquad 6.27$$

[6]The equivalent shock frequency and the natural frequency of the product on its cushion are numerically equal.

During the impact the amplification factor is given by the expression:

$$A_m (0 \leqslant t \leqslant \tau) = \frac{\frac{f_1}{f_2}}{(\frac{f_1}{f_2} - 1)} \sin \frac{2N\pi}{(\frac{f_1}{f_2} + 1)} \qquad 6.28$$

where N is an integer, f_2 the equivalent shock frequency and f_1 the natural frequency of the critical element[7]. Just after impact the amplification factor is:

$$A_m (t > \tau) = \frac{2 (\frac{f_1}{f_2}) \cos (\frac{f_1}{2f_2})}{1 - (\frac{f_1}{f_2})^2} \qquad 6.29$$

The largest value of A_m calculated from Equations 6.28 and 6.29 must be chosen for Equation 6.27. This depends on the relationship of f_1 and f_2:

1. If $f_1/f_2 < 1$, $f_1 < f_2$ or $1/2\,T_1 > 1/2\,T_2 = \tau$, i.e. if the duration of the shock is less than one half the CE's natural period of vibration, the maximum amplification occurs just after the impact is over, and,

   $A_m = A_m (t > \tau)$ (Equation 6.29).

2. If $f_1/f_2 > 1$, $f_1 > f_2$ or $1/2\,T_1 < 1/2\,T_2 = \tau$, i.e. if the duration of the shock is longer than one half the CE's natural period of vibration, the maximum amplification occurs during the shock, and,

   $A_m = A_m (0 \leqslant t \leqslant \tau)$ (Equation 6.28).

The maximum amplification values are given in Figure 6.11 for frequency ratios between 0.01 and 11.0. Notice that as the f_1/f_2 ratio becomes large the value for A_m approaches 1, reflecting the "direct" transfer of the shock to a stiff element. For small values of the frequency ratio:

$$A_m = 2 (\frac{f_1}{f_2}), \quad \text{for } \frac{f_1}{f_2} \leqslant .20$$

Let's take a simple example. Suppose we have a product which contains a critical element, as shown in Figure 6.12. The natural frequency of vibration for this critical element is 38 Hz. A shock pulse is applied to the product. The shock pulse may be approximated by a half sine, the peak acceleration is 100 g's and the duration is 25 milliseconds. What is the shock to the critical element?

---

[7] $f_1 = \frac{1}{2\pi}\sqrt{\frac{k_1 g}{W_1}}$. This frequency is usually determined experimentally by vibrating the product and observing the resonant frequency of the critical element.

f_1/f_2	A_m	f_1/f_2	A_m	f_1/f_2	A_m	f_1/f_2	A_m	f_1/f_2	A_m	f_1/f_2	A_m
.01	.020	.82	1.397	1.66	1.768	2.50	1.625	3.85	1.300	6.90	1.169
.02	.040	.84	1.419	1.68	1.767	2.52	1.620	3.90	1.289	7.00	1.167
.03	.060	.86	1.441	1.70	1.767	2.54	1.615	3.95	1.279	7.10	1.164
.04	.080	.88	1.462	1.72	1.765	2.56	1.610	4.00	1.268	7.20	1.160
.06	.120	.90	1.482	1.74	1.764	2.58	1.605	4.05	1.258	7.30	1.157
.08	.160	.92	1.501	1.76	1.762	2.60	1.600	4.10	1.247	7.40	1.153
.10	.200	.94	1.520	1.78	1.761	2.62	1.595	4.15	1.237	7.50	1.149
.12	.239	.96	1.538	1.80	1.759	2.64	1.590	4.20	1.227	7.60	1.145
.14	.279	.98	1.555	1.82	1.757	2.66	1.585	4.25	1.217	7.70	1.140
.16	.318	1.00	1.571	1.84	1.755	2.68	1.580	4.30	1.207	7.80	1.135
.18	.357	1.02	1.586	1.86	1.753	2.70	1.575	4.35	1.198	7.90	1.131
.20	.396	1.04	1.601	1.88	1.750	2.72	1.570	4.40	1.188	8.00	1.126
.22	.435	1.06	1.614	1.90	1.747	2.74	1.565	4.45	1.179	8.10	1.120
.24	.474	1.08	1.627	1.92	1.745	2.76	1.560	4.50	1.170	8.20	1.115
.26	.512	1.10	1.640	1.94	1.742	2.78	1.555	4.55	1.160	8.30	1.110
.28	.550	1.12	1.651	1.96	1.739	2.80	1.550	4.60	1.151	8.40	1.104
.30	.588	1.14	1.662	1.98	1.735	2.82	1.545	4.65	1.142	8.50	1.099
.32	.625	1.16	1.672	2.00	1.732	2.84	1.540	4.70	1.133	8.60	1.093
.34	.662	1.18	1.682	2.02	1.729	2.86	1.535	4.75	1.125	8.70	1.087
.36	.698	1.20	1.690	2.04	1.725	2.88	1.530	4.80	1.116	8.80	1.082
.38	.735	1.22	1.699	2.06	1.722	2.90	1.525	4.85	1.108	8.90	1.076
.40	.771	1.24	1.706	2.08	1.718	2.92	1.520	4.90	1.099	9.0	1.070
.42	.806	1.26	1.714	2.10	1.714	2.94	1.515	4.95	1.091	9.1	1.075
.44	.841	1.28	1.720	2.12	1.710	2.96	1.510	5.00	1.083	9.2	1.080
.46	.875	1.30	1.726	2.14	1.706	2.98	1.505	5.10	1.098	9.3	1.083
.48	.909	1.32	1.732	2.16	1.702	3.00	1.500	5.20	1.112	9.4	1.087
.50	.943	1.34	1.737	2.18	1.698	3.05	1.488	5.30	1.124	9.5	1.090
.52	.976	1.36	1.742	2.20	1.694	3.10	1.475	5.40	1.134	9.6	1.092
.54	1.008	1.38	1.746	2.22	1.690	3.15	1.463	5.50	1.143	9.7	1.094
.56	1.040	1.40	1.750	2.24	1.685	3.20	1.451	5.60	1.150	9.8	1.097
.58	1.071	1.42	1.753	2.26	1.681	3.25	1.438	5.70	1.157	9.9	1.098
.60	1.102	1.44	1.757	2.28	1.676	3.30	1.426	5.80	1.162	10.0	1.100
.62	1.132	1.46	1.759	2.30	1.672	3.35	1.414	5.90	1.167	10.1	1.101
.64	1.162	1.48	1.761	2.32	1.667	3.40	1.402	6.0	1.170	10.2	1.102
.66	1.191	1.50	1.763	2.34	1.663	3.45	1.390	6.10	1.172	10.3	1.102
.68	1.219	1.52	1.765	2.36	1.658	3.50	1.379	6.20	1.174	10.4	1.103
.70	1.246	1.54	1.766	2.38	1.654	3.55	1.367	6.30	1.175	10.5	1.103
.72	1.273	1.56	1.767	2.40	1.649	3.60	1.356	6.40	1.176	10.6	1.103
.74	1.299	1.58	1.768	2.42	1.644	3.65	1.344	6.50	1.175	10.7	1.102
.76	1.325	1.60	1.768	2.44	1.639	3.70	1.333	6.60	1.175	10.8	1.102
.78	1.349	1.62	1.769	2.46	1.635	3.75	1.322	6.70	1.173	10.9	1.101
.80	1.373	1.64	1.768	2.48	1.630	3.80	1.311	6.80	1.172	11.0	1.100

Figure 6.11. Amplification Factors (Half Sine Pulse)

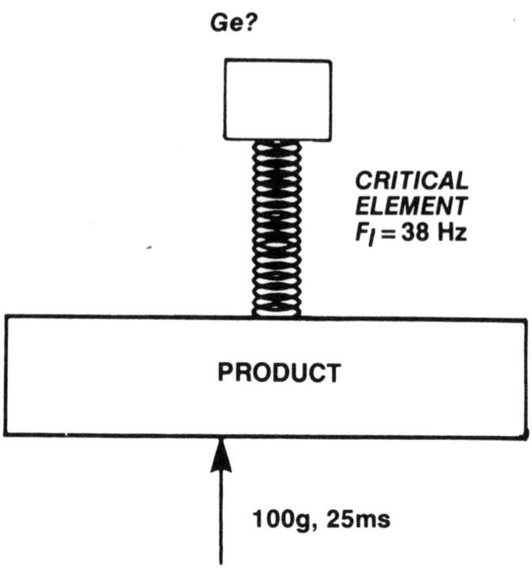

Figure 6.12. An Example of Shock Amplification.

Solution:

1. $\tau = \dfrac{25 \text{ milliseconds}}{1000 \text{ milliseconds/second}} = 0.0250 \text{ seconds}$

2. $f_2 = 1/2\tau = 1/2\ (.0250 \text{ sec}) = 20 \text{ Hz}$

3. $\dfrac{f_1}{f_2} = \dfrac{38 \text{ Hz}}{20 \text{ Hz}} = 1.90$

4. $\dfrac{f_1}{f_2} = 1.90$ corresponds to an A_m (from chart) of 1.747;

   $A_m = 1.747$

5. $G_e = A_m \times G_m = 1.747 \times 100 \text{ g's} = 174.7 \text{ g's}$

We see that the critical element experienced a shock of about 175 g's as a result of the 100 g shock to the product.

In order to predict damage or evaluate the cushion effectiveness we need to know the highest acceleration level a critical element can withstand without damage. This is defined as G_s, the **safe** level; the maximum value of G_e to occur without damage to the element. Shock accelerations at the CE exceeding G_s will result in damage. An example will illustrate this approach.

Suppose using a shock machine (Figure 6.13) programmed to produce half sine shock pulses of 20 milliseconds duration, we find that a product can withstand 100 g's. At higher accelerations there is damage to a critical element. The critical element has a natural frequency of 80 Hz and the entire product weighs four pounds. In distribution we will be using a cushion for which $k_2 = 750$ lb./in. and we expect a maximum drop height of four feet. Will we be safe? First, we must determine G_s using test data:

1. Calculate the equivalent shock frequency of the shock machine:

   $\tau = 20 \text{ ms} = 0.020 \text{ sec}$

   $f_2 = \dfrac{1}{2\ (.020 \text{ sec})} = 25 \text{ Hz}$

2. Find the ratio of the element's f_1 to the shock machine's f_2:

   $\dfrac{f_1}{f_2} = \dfrac{80}{25} = 3.2$

   Then look up the amplification value from Figure 6.11

   $A_m = 1.451$

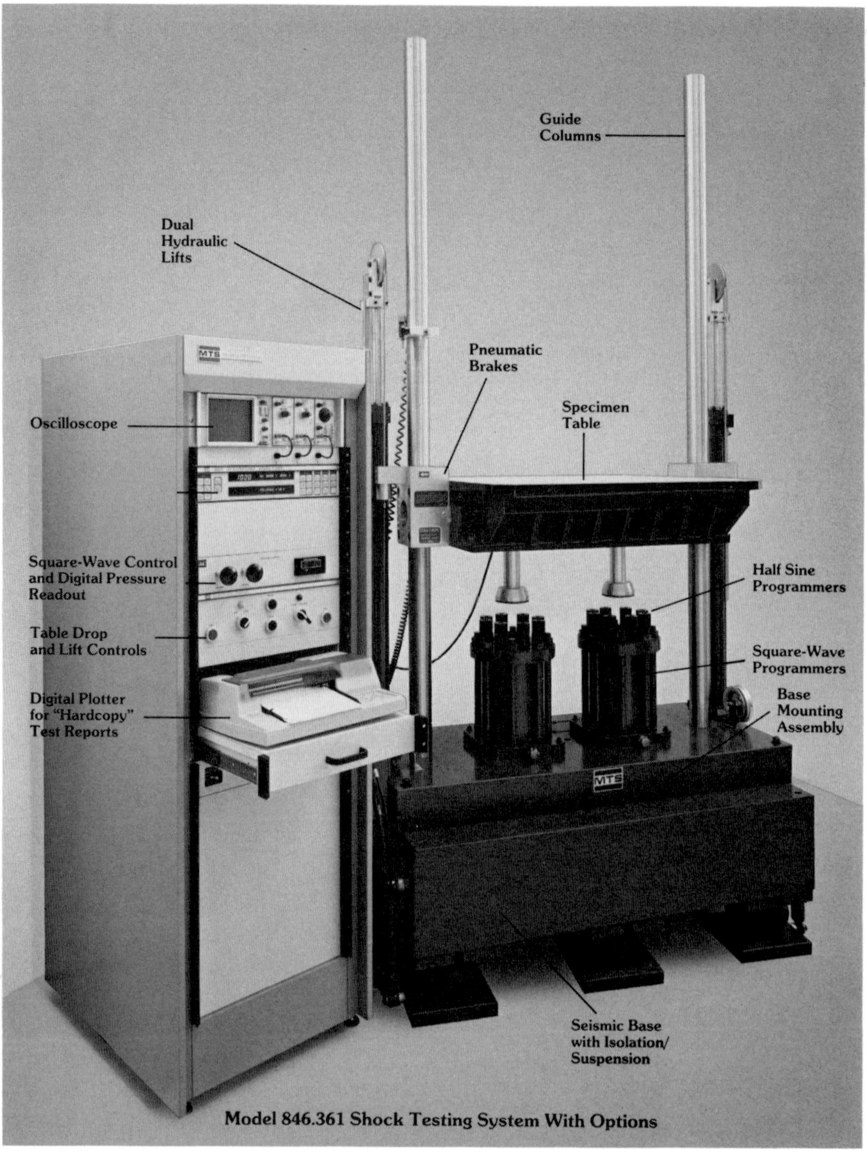

Figure 6.13. A Shock Machine.

3. The 100 g's (G_m) produced by the shock machine produces a response at the CE of:

$$G_e = A_m G_m = (1.451)(100 \text{ g}) = 145.1 \text{ g's}$$

Therefore, $G_s = 145.1$ g's.

Now we look at what will happen in distribution.

4. The shock in a 4 foot drop is:

$$G_m = \sqrt{\frac{2 k_2 h}{W_2}} = \sqrt{\frac{2 \cdot 750 \text{ lb/in} \cdot 48 \text{ in}}{4 \text{ lb.}}}$$

$$G_m = 134 \text{ g's}$$

5. The shock duration on the cushion is:

$$\tau = \frac{\pi}{\sqrt{\frac{k_2 g}{W_2}}} = \frac{\pi}{\sqrt{\frac{750 \cdot 386.4}{4}}} = 0.0117 \text{ sec}$$

Then

$$f_2 = \frac{1}{2\tau} = 42.83 \text{ Hz or} \sim 43 \text{ Hz}$$

6. $\frac{f_1}{f_2} = \frac{80}{43} = 1.87$, and from Figure 6.11, $A_m = 1.752$

7. So the shock at the CE during distribution is:

$$G_e = (1.752)(134 \text{ g}) = 234.8 \text{ g's}$$

$$G_e = 234.8 \text{ g's} > 145.1 \text{ g's} = G_s$$

Because the G_e in distribution will be larger than the safe level, damage will occur and we may conclude that this cushion is inadequate to protect the product in a 4 foot drop.

The next example will illustrate how shock amplification theory may be used to calibrate and interpret a simple spring-mass accelerometer like the one shown in Figure 6.14.

The accelerometer measuring stylus has a natural frequency of 18 Hz and behaves like a critical element. The accelerometer was calibrated by securing it to a shock machine, generating a 10 ms shock and observing the response of the stylus. For every 10 g's at the shock machine table, the stylus registered one division, giving us an **apparent** calibration of 10 g's/division. When the mechanical accelerometer was used in the field a 36 g, 40 ms shock was encountered. What reading did the mechanical accelerometer give in response to this shock?

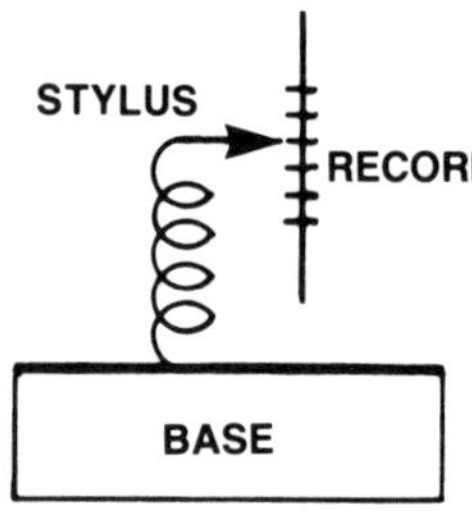

Figure 6.14. A Spring-Mass Accelerometer.

First, we need to examine the calibration of the mechanical accelerometer. In the calibration process, a shock of 10 ms duration was used. The equivalent shock frequency is:

$$f_2 = \frac{1}{2(.010\ \text{sec})} = 50\ \text{Hz}$$

The frequency ratio was then

$$\frac{f_1}{f_2} = \frac{18}{50} = 0.36$$

From Figure 6.11,

$$A_m = 0.698$$

The problem states that for every 10 g's input the stylus registered one division. Therefore, if a 100 g shock was produced at the table, the stylus would respond with a 10 division movement. But we know for such a shock the acceleration at the stylus (or element) was only,

$$G_e = A_m G_m = (0.698)(100\ \text{g's}) = 69.8\ \text{g's}$$

It was this 69.8 g shock at the stylus that caused it to register 10 divisions, so the **real** calibration, using the actual shock to the stylus is:

$$\frac{69.8\ \text{g}}{10\ \text{div}} = 6.98\ \text{g's/div},$$

Not 10 g/div as we originally guessed.

This real calibration may also be calculated by multiplying the amplification factor by the apparent calibration:

$$(0.698)\,(10\ g/div) = 6.98\ g/div$$

In use, this accelerometer encountered a 36 g, 40 ms shock. The equivalent shock frequency here is,

$$f_2 = \frac{1}{2(.040\ sec)} = 12.5\ Hz$$

The frequency ratio was then,

$$\frac{f_1}{f_2} = \frac{18}{12.5} = 1.44,$$

corresponding to an amplification factor of:

$$A_m = 1.757$$

The 36 g shock is amplified at the recording stylus,

$$G_e = (1.757)\,(36\ g) = 63.3\ g$$

The 63.3 g's at the stylus causes it to move;

$$\frac{63.3\ g}{6.98\ g/div} = 9.1\ div.$$

The 36 g, 40 ms shock pulse causes a respons of 9.1 divisions at the recording stylus (not the 3.6 divisions you would have guessed if you knew nothing of shock amplification).

6.5 Horizontal Impacts

Up to this point we have dealt strictly with vertical impacts. But what happens when we are confronted with a horizontal impact, as in this example?

"A 40 pound product is encased in linear cushioning material, for which k = 200 lb/in. The product within its cushioning is securely fastened to the floor of a railcar. The railcar is coupled to the rest of the train at **10 miles per hour.** If the working length of the cushion materials is 40% of the total length, what is the minimum thickness of the cushion needed to prevent "bottoming out" against the outside container? What are G_m and τ at the product?"

We may solve this problem by computing a vertical drop height which produces the same impact velocity as the horizontal impact.

$$v_I = 10 \text{ miles/hr} = \frac{10 \text{ miles} \cdot 5280 \text{ ft/mile} \cdot 12 \text{ in/ft}}{3600 \text{ sec/hr}} = 176 \text{ in/sec}$$

$$v_I = \sqrt{2\,gh}$$

$$h = \frac{v_I^{\,2}}{2\,g} = \frac{(176 \text{ in/sec})^2}{2 \cdot 386.4 \text{ in/sec}^2} = 40 \text{ in.} \qquad 6.30$$

Now we may compute d_m, G_m and τ as if we had a vertical drop from 40 inches:

$$d_m = \sqrt{\frac{2\,W_2 h}{k_2}} = \sqrt{\frac{2 \cdot 40 \text{ lb.} \cdot 40 \text{ in.}}{200 \text{ lb/in}}} = 4 \text{ in.}$$

$$G_m = \sqrt{\frac{2 k_2 h}{W_2}} = \sqrt{\frac{2 \cdot 200 \text{ lb/in} \cdot 40 \text{ in}}{40 \text{ lb}}} = 20 \text{ g}$$

$$\tau = \pi \cdot \sqrt{\frac{W_2}{k_2 g}} = \pi\sqrt{\frac{40 \text{ lb}}{200 \text{ lb/in} \cdot 386.4 \text{ in/sec}^2}} = 0.071 \text{ sec}$$[8]

Since WL = 0.40 TT, we may compute the required cushion thickness:

$$d_m = WL = 0.40\ TT$$

$$TT = \frac{d_m}{0.4} = \frac{4 \text{ in}}{0.4} = 10 \text{ in.}$$

Some useful expressions may also be generated for G_m and d_m following the reasoning above.

$$G_m = \sqrt{\frac{2\,k_2 h}{W_2}}, \text{ and } h = \frac{v_I^{\,2}}{2\,g} \text{ from Equation 6.30.}$$

Substituting for h,

$$G_m = \sqrt{\frac{2\,k_2}{W_2} \cdot \frac{v_I^{\,2}}{2\,g}} = v_I\sqrt{\frac{k_2}{W_2 g}} \qquad 6.31$$

From $\delta_{st} = \frac{W_2}{k_2}$, Equations 6.31 becomes

$$G_m = \frac{v_I}{\sqrt{\delta_{st}\, g}} \qquad 6.32$$

---

[8]We have assumed here that only one of the two side cushions is acting on the product. If both side cushions act, staying in contact with the product during impact, the expression for τ is:

$$\tau = \pi\sqrt{\frac{W_2}{2\,k_2 g}}$$

The natural frequency is given by

$$f_2 = 3.13 \sqrt{1/\delta_{st}}, \ \delta_{st} \text{ in inches}$$

so

$$G_m = \frac{v_I f_2}{3.13 \sqrt{g}} = \frac{v_I f_2}{61.53} \text{ (if } v_I \text{ and } f_2 \text{ are in in/sec and Hz)} \tag{6.33}$$

Likewise, an expression for d_m may also be generated;

$$d_m = \sqrt{\frac{2 W_2 h}{k_2}} = \sqrt{\frac{2W v_I^2}{k_2 \, 2 g}} = v_I \sqrt{\frac{W_2}{k_2 g}} = v_I \sqrt{\frac{\delta_{st}}{g}}$$

$$= \frac{v_I}{\sqrt{g}} \frac{3.13}{f_2} = 0.16 \frac{v_I}{f_2} \tag{6.34}$$

PROBLEM SET, CHAPTER 6

1. An article weighing 12 lbs. can sustain a maximum shock of 80 g's (G_m = 80). Find k_2 for a linear cushion which will limit the acceleration of the product to 80 g's in a 36 inch drop. What is the maximum deflection of the cushion? Find the duration (τ) of the shock pulse sustained by the article.

2. An item weighing 500 lbs. can sustain a maximum shock of 5 g's. Assuming a drop height of 24 inches, compute k_2 for a linear cushion. Find the maximum deflection of the spring and compute the duration of the shock pulse (τ).

3. If the item in Problem 2 measures 12" x 12" x 12" and is equally sensitive to shock when dropped on any of its six (6) faces, and the working length of the springs used is half of the total length of the spring, compute the interior dimensions of the package necessary to contain the system.

4. An item which is damaged at G_m over 40 is cushioned with linear springs which have a working length of 5/8 of their total length. Find the dimensions of the box needed to hold the item and the required cushioning if it is equally sensitive to shock in all directions and if the item has dimensions of 11" x 14" x 17". The maximum drop height expected in distribution is 42 inches.

5. Find the dimensions of a box needed to hold an item measuring 3" x 6" x 2.5" with an acceleration sensitivity of 25 and the required linear springs to protect it if it is dropped 18" and the working length of the springs is 7/8 of the total length. Assume that the item is equally sensitive to shock in all directions.

6. An article which weighs 18 pounds is supported on a linear spring with a spring constant k = 882 pounds per inch. If the item is dropped 30 inches, find the number of g's it encounters and the duration of the shock pulse.

7. An article which weighs 19 lbs. is supported on a linear spring with a spring constant k = 931 lbs/in. If the item is dropped 24 inches, find the number of g's it encounters and the duration of the shock.

8. An item measures 4" x 6" x 8". For impacts on either 4" x 6" face, A_c* = 25. For impacts on either 4" x 8" face, A_c = 42. For impacts on either 6" x 8" face, A_c= 18. Find the dimensions of the box needed to contain the item and its protective linear spring cushions if the maximum drop height is 38", and if the working length of the springs is 50% of the total length.

9. You have an instrument for measuring shocks applied to articles. This instrument weighs three pounds and consists of three spring-mass systems mounted at right angles to one another to record shocks along three orthogonal axes. When rigidly mounted to a larger article, the instrument effectively becomes three elements of the larger article. Each spring-mass system has a natural frequency of 25 Hz.

---

*A_c = maximum tolerable acceleration.

The recorder has been calibrated at 5 Hz to read 10 g's per division. (This means that the person doing the calibration gave the article a known acceleration input at 5 Hz and measured the instrument's response in divisions — the apparent calibration is the ratio of input g's to recorder divisions.) What is the true calibration?

A. Find the recorder reading for a 100 g, 25 ms shock pulse along any axis.

B. Find the recorder reading for a 100 g, 10 ms shock pulse along any axis.

C. Find the recorder reading for a 25 g, 10 ms shock pulse along any axis.

D. Find the recorder reading for a 50 g, 5 ms shock pulse along any axis.

10. The recorder in Problem 9 is rigidly attached to a 40 lb. table. (For purposes of this problem we will consider the recorder the lowest natural frequency element of the table.) The table and its package which weighs 10 lbs. are dropped 24 inches. The recorder reads five divisions. What do you know about the shock?

    The recorder in Problem 9 is impacted so that all three elements move. Element one reads five divisions, element two reads seven divisions, and element three reads three divisions. What can you say about the shock?

11. A simple spring mass accelerometer with an element natural frequency of 18 Hz has been calibrated using a shock machine which produces 10 millisecond duration pulses to read 10 g's per division. In use a shock of 36 g's having a duration of 40 milliseconds is encountered. What reading will the instrument show?

12. A simple spring mass accelerometer ($f_n = 18$ Hz) is subjected to a shock of 18 g's which has a duration of 60 milliseconds. What reading would you expect, knowing that the accelerometer had been calibrated using a shock machine to apparently register 23 g's per division with a shock duration of 8 milliseconds?

13. An article weighing 27.5 lbs. can sustain a maximum shock of 52 g's. Find the spring constant for a linear, undamped cushion which will limit the acceleration on the article to 52 g's in a five foot drop. What is the maximum deflection of the cushion? What is the duration of the shock pulse sustained by the article?

14. A. Find G_m in a situation where a 147 lb. article is dropped 38 inches onto a cushion with a linear spring $k_2 = 1462$ lb/in. What is the duration of the shock?

    B. What is the difference between the shocks produced (G_m, τ) when an article is dropped a distance h and then a distance of 2h? (Compare the two shocks produced at drop heights h and 2h).

15. Using a shock machine programmed to produce a half-sine shock pulse of 20 milliseconds duration, a 5 lb. product is tested (without cushioning). It is found that the product is undamaged at 100 g's, but is damaged at 101 g's and above.

The damage occurred to a critical element; the natural frequency of this critical element is 60 Hz. The product is packaged with linear springs for which $k_2 =$ 1010 lb/in. A maximum drop height of 60 inches is expected in the distribution system.

A. What is the frequency of the shock produced by the shock machine?

B. What is the amplification factor for the test situation?

C. What is G_s, the safe acceleration level for the critical element?

D. What is the magnitude of the acceleration experienced by the product in the distribution system?

E. What is the duration of the shock experienced by the product in the distribution system?

F. What is the amplification factor in the distribution system?

G. Does the cushion material provide adequate protection for the product (and critical element)?

16. A 10 lb. product has two critical elements. The natural frequency of critical element A is 22 Hz. The natural frequency of critical element B is 38 Hz. The product is dropped from a height of three feet onto a linear undamped cushion for which k = 500 lb/in.

A. What is the amplitude (in g's) of the shock experienced by critical element A?

B. What is the amplitude (in g's) of the shock experienced by critical element B?

17. A 25 lb. product measuring 6" x 6" x 8" can withstand a maximum shock of 55 g's on any of its six faces. Compute the linear spring constant of the springs necessary to protect the product in a 36 in. fall. What will be the maximum deflection of the linear springs? If the working length of each spring is 65% of its total length, find the inner dimensions of the container needed to hold the item and the cushion springs.

18. An element of an article has a natural frequency of 35 Hz. A shock machine with plastic programmers which produce 2 ms shock pulses was used to test the article. The element survived a shock of 115 g's applied to the article but broke when the article received a shock of 120 g's. What is the maximum safe shock on the element?

19. Using a shock machine which produces a shock pulse of 0.015 seconds, you have calibrated a simple spring-mass accelerometer. In the calibration procedure you find that 100 g's moves the indicator stylus 10 divisions. The accelerometer has a natural frequency of 25 Hz. In a test situation the

accelerometer receives a shock of 85 g's with a duration of 0.007 seconds. What reading, in divisions, will the accelerometer recorder show?

20. If the product-cushion system in Problem 7, Chapter 4, is dropped cushion side down from a height of 28 inches, what shock will be experienced by the product?

21. An article has an element with a natural frequency of 40 Hz. The element will break if it receives a shock of more than 37 g's. Will the article survive a 42 g, 5 ms shock?

22. A product weighs 30 lbs. and measures 5" x 8" x 6". In testing, we find that it can withstand up to 25 g's applied in any direction. If we expect a maximum drop height of 3 feet, and drops on any face:

    A. What is the spring constant, k_2, of the linear cushion necessary to protect this product?

    Rather than use a cushion with the k_2 found in Part A, we select cushioning material with a $k_2 = 200$ lb/in, for which the working length is equal to 75% of the total length.

    B. Will this cushion material adequately protect our product in a 3 foot drop?

    C. What is the minimum thickness of each of the six cushions?

    D. What are the inside dimensions of the container which will hold the product and its cushions?

    E. What is the duration of the shock when the product-cushion is dropped 3 feet?

23. A ten pound product is cushioned as shown in the diagram below:

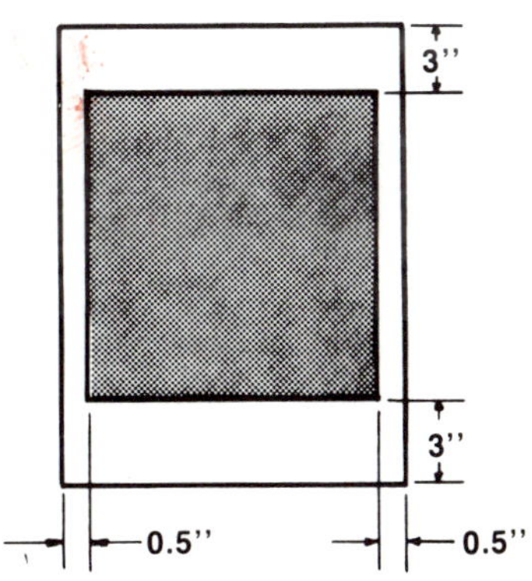

The linear spring constant for all the material is 500 lb/in. The product has been tested and it was found that the maximum G_m along the vertical axis should not exceed 50 g's, and that the maximum G_m along the horizontal axis should not exceed 25 g's. The cushion material used has a working length equal to 60% of its total length.

A. Is the product adequately protected (in terms of maximum shock) if it is dropped on its bottom from a vertical height of 42 inches?

B. Is there enough cushion present to prevent bottoming out?

C. Is the product adequately protected for a horizontal impact on its side at 325 feet per minute?

D. Is there enough cushion present to prevent contact (horizontal bottoming out) between the product and outside container during impact?

24. An assembly weighing five pounds has one spring-mass system comprising an element of the assembly. This element has a natural frequency of 60 Hz. The assembly survived a 150 g, 11 ms shock in the shock machine, but did not survive a 160 g, 11 ms shock (the element failed). Design linear cushioning to protect this item in a 30 inch drop. (Design your cushion so that d_m is minimized by maximizing k_2.)

25. A small, one pound mechanical accelerometer is calibrated in the laboratory using the shock machine. The shock machine is set to produce half sine pulses of 5 ms duration. The recording stylus of the mechanical accelerometer has a natural frequency of 25 Hz. It is found that for a shock of 200 g's, the recording stylus registers 19.6 divisions.

A. What is the actual calibration of this mechanical accelerometer?

A product is tested for fragility in the laboratory using the shock machine. This product weighs 50 lbs. and contains a small lightweight critical element which has a natural frequency of 60 Hz. The product is securely fastened directly on the shock machine. The shock machine is programmed to produce half sine pulses of 25 ms duration. It is found that the product remains undamaged at 120 g's, but damage to the critical element occurs at 121 g's.

B. What is G_s, the critical acceleration for the product's critical element?

The mechanical accelerometer described in Part A is securely fastened to the product described in Part B. (Note that the product and the accelerometer together weigh 51 pounds.) They are then encased on all sides in 6 inches of linear cushioning material for which $k = 2084$ lb/in. The package is sent through the distribution system and then opened. The accelerometer shows a reading of 34 divisions.

C. Based on the reading of the mechanical accelerometer, would you expect the product to be damaged? Explain. (Hint: You know k, so you can compute τ in the distribution system.)

D. If the working length of the cushion material is 50% of the total length, was there enough cushion material present to prevent bottoming out?

26. A simple spring mass accelerometer has a recording element with a natural frequency of 15 Hz. Using a shock machine which produces 8 ms shock pulses the accelerometer was calibrated to apparently read 15 g's per division.

   A. When using the accelerometer, a 50 g shock of 28 ms duration was encountered. What was the reading (in divisions) registered by the accelerometer?

   B. In another situation, the accelerometer registers 15 divisions for a shock that was measured to have a frequency of 43 Hz. What was the magnitude of this shock in g's?

27. A 12 lb. product sits upon a cushion. Its static deflection is observed to be 0.08 inches. This product has a critical element within it, having a natural frequency of 4.43 Hz. This element will break if it receives more than 25 g's ($G_S = 25$ g's). If the product is dropped from a height of 36 inches, will it be damaged?

28. An element of a 60 lb. article has a natural frequency of 26 Hz. The article was attached firmly to a shock machine outfitted with 2 ms programmers and tested for damage. It was found that with the 2 ms shocks, the article was undamaged at 200 g's, but damaged at 205 g's. What is the maximum safe shock on the element?

29. The article in Problem 28 is removed from the shock machine and cushioned on all sides with 7 inches of linear cushion material for which k = 500 lb/in. The working length of this cushion is 75% of the total thickness. The article with its cushioning is dropped 92 inches. Is the article adequately protected to survive this drop without damage? Is there enough cushioning material present to prevent bottoming out?

30. A 24 lb. product was dropped from 40 inches onto a stack of N identical cushions. A shock (G_m) of 31.62 g's was recorded at the product. Each cushion has a spring constant of 2100 lb/in. How many cushions were in the stack?

7. SHOCK IN DISTRIBUTION, PRODUCT FRAGILITY AND CUSHION DESIGN

7.0 Mechanical Shock in the Distribution System

In the last chapter we saw that two of the parameters related to product damage could be approximated by:

$$G_m = \sqrt{\frac{2\,k_2 h}{W_2}} \qquad \text{(Equation 6.15)}$$

$$\tau = \frac{\pi}{\sqrt{\frac{k_2 g}{W_2}}} \qquad \text{(Equation 6.24)}$$

The maximum acceleration level at the product G_m is related to the drop height (h), product weight (W_2), and the combined cushion characteristics of the cushion, container and product (k_2)[1]. The shock duration was related to the product weight and the cushion characteristics. To define the shock environment during distribution we need to know something of the drop heights encountered by our product-package system as well as the system's cushion characteristics.

Unfortunately, very little is known about the drop heights and shock levels encountered in distribution, except that every distribution system has a unique, complex profile. Table 7.1 contains some generalizations linking drop height, handling and package characteristics. Notice that as the package becomes larger, either in weight or physical size, the expected drop height decreases. As you might expect, the more cumbersome a package becomes the less likely it is to be raised to a great height, either by man or machine, and therefore the less likely it is to be dropped from a great height. This is partially illustrated by the results of an empirical study(12), shown in Figures 7.1, 7.2, and 7.3. Here it appears that the 25 lb. package is more likely to be dropped from any height between five and ninety inches than is a 43 lb. package. However, from Figure 7.3, it is also likely that even the heavier package will be dropped from a relatively great height, although any package will more likely receive numerous drops from relatively low drop heights.

Figure 7.4 provides some information on the horizontal impacts (see Section 6.5) involved in railcar coupling. Here, the horizontal impact velocity must be greater than two miles per hour for the automatic couplers to function correctly. Notice that horizontal impacts of ten and eleven miles per hour are not uncommon.

Figure 7.5 contains information on the probability of drop height as a function of package weight. Such distributions are only of limited usefulness, however, as they may not reflect the peculiarities of a particular distribution system. The careful packaging technologist will want to personally investigate the specific distribution system of interest, if it is at all possible, rather than rely on such generalizations.

---

[1] We assumed that all the cushioning was done by the linear spring, but the assignment of partial cushioning to the product and container amounts to much the same thing.

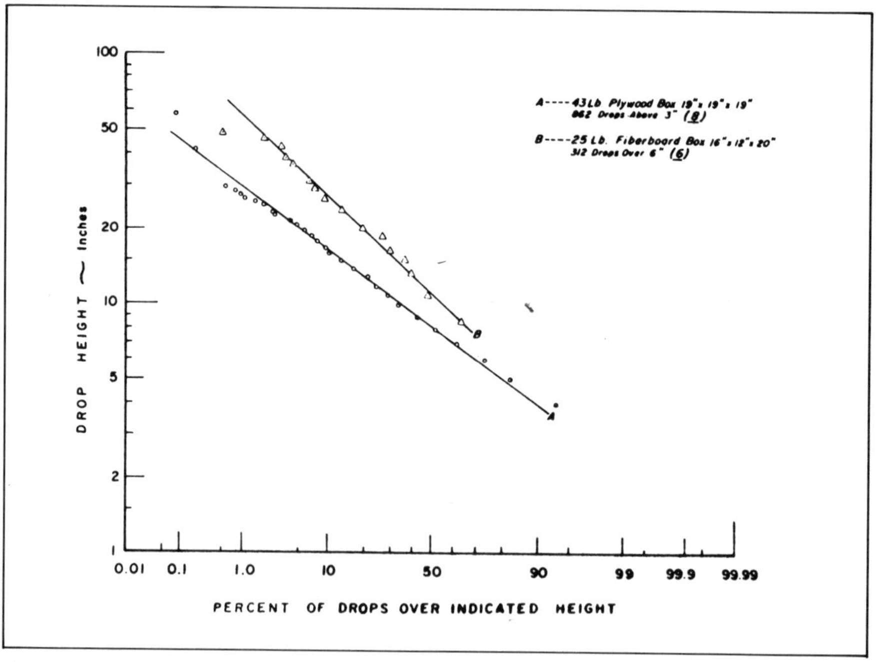

Figure 7.1. Drop Height versus Probability of Occurrence. (12)

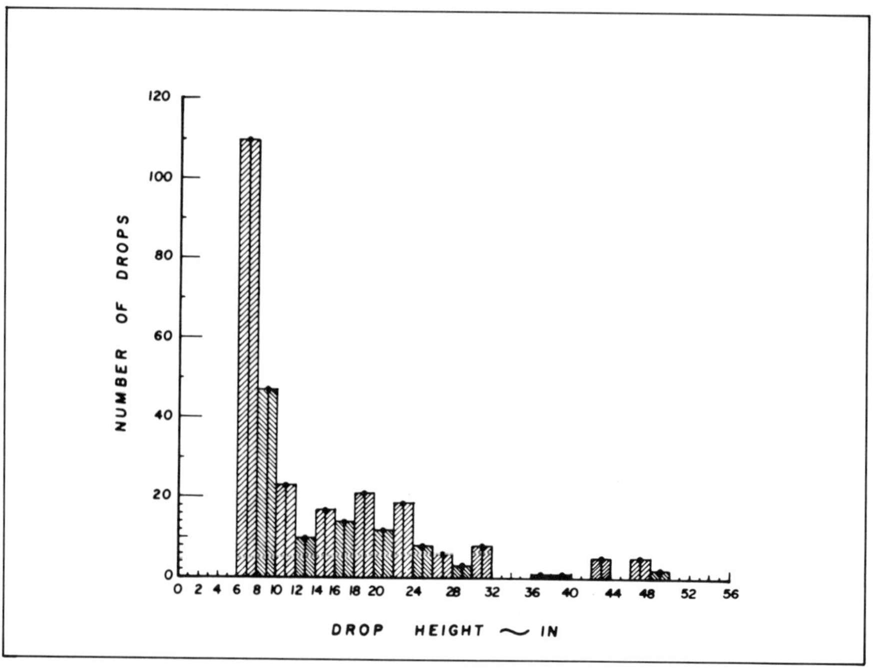

Figure 7.2. Drop Height versus Number of Drops for a 25 lb. Package. (12)

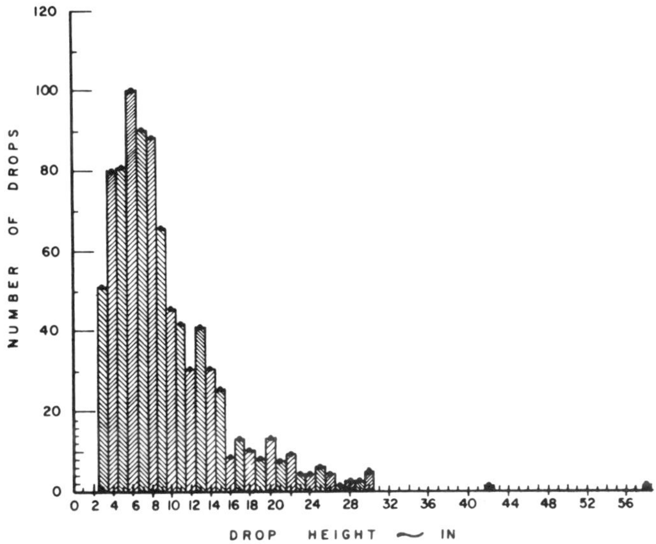

Figure 7.3. Drop Height versus Number of Drops for a 43 lb. Package. (12)

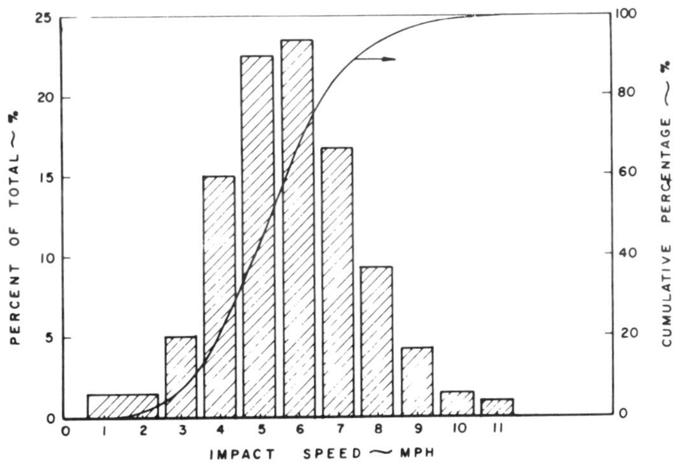

Figure 7.4. Distribution of Impact Speeds Involved in Railcar Coupling. (12)

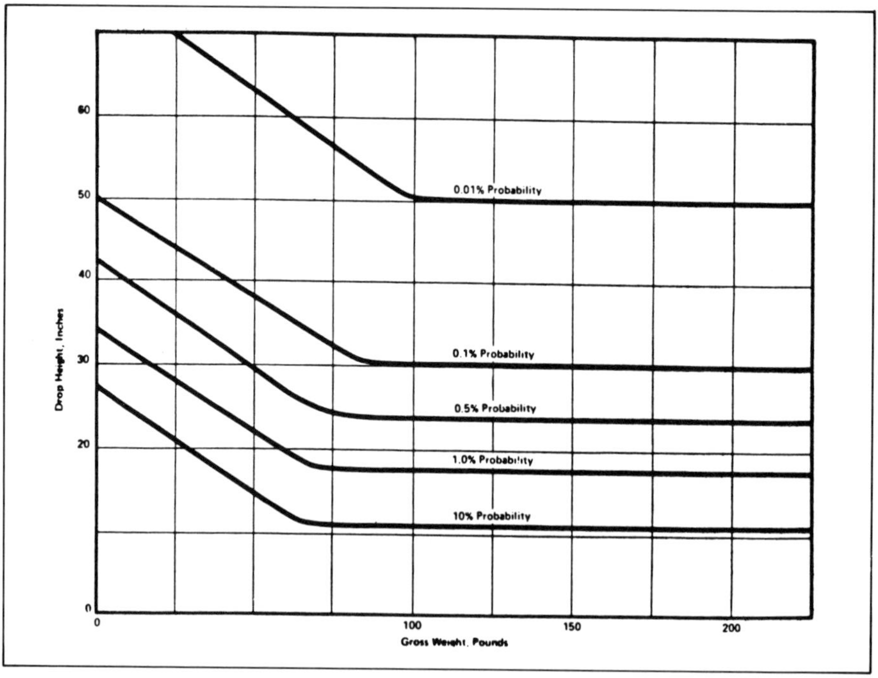

Figure 7.5. Probability Curves for Handling Shocks. (22)

TABLE 7.1
HANDLING ENVIRONMENTS

Severest Probable Environment				
Package Weight (pounds)	**Greatest Dimension (inches)**	**Drop (inches)**	**Form of Type of Drop**	**Handling**
20	48	42	Any side or corner	One man throw
20-50	36	36	Any side or corner	One man carry
50-100	48	24	Any side or corner	Two men carry
100-150	60	21	Any side or corner	Two men carry
150-200	60	18	Any side or corner	Two men carry
200-600 Mechanical	72	24	Rotating, either end-roll or tip	
600-3,000 Mechanical	Unlimited	18	Rotating, either end-roll or tip	
3,000 Mechanical	Unlimited	12	Rotating, either end-roll or tip	

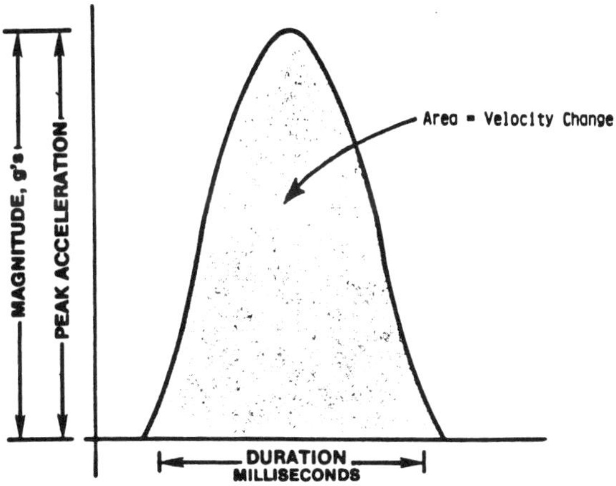

Figure 7.6. The Shock Pulse.

7.1 The Damage Boundary Curve

Table 7.2 contains typical published information on the **fragility** of articles. We know, however, from the discussion on shock amplification in the last chapter that the shocks delivered to components within an article depend on the durations of the shock pulses and the natural frequencies of the components. Thus, damage to the article is dependent not only on the maximum shock level at the article, but also on the shock duration. A true product fragility index must be based on two of the three variables related to shock damage:

Maximum shock amplitude, G_m,

Pulse duration, τ,

Velocity change, Δv,

where velocity change is numerically equal to the area under the acceleration versus time graph of the shock pulse. Referring to the half sine pulse shown in Figure 7.6, the velocity change is:

$$\Delta v = \frac{2}{\pi}(\text{peak acceleration, in/sec}^2)\ (\text{duration, sec}) \quad \text{or}$$

$$\Delta v = \frac{2}{\pi} G_m \cdot g \cdot \tau \qquad 7.1$$

where $\frac{2}{\pi}$ is the "shape factor" relating the shape of the half sine pulse to its area[2]. If we know the shape of a shock pulse, we need only to know two of the three damage parameters (G_m, τ and Δv).

TABLE 7.2
APPROXIMATE FRAGILITY OF TYPICAL PACKAGED ARTICLES*

Extremely Fragile	
Missile guidance systems, precision aligned test instruments	15-25 g's
Very Delicate	
Mechanically shock mounted instruments and electronic equipment (shock mounts should be firmly secured prior to packaging. They are provided for in-service protection only)	25-40 g's
Delicate	
Aircraft accessories, electric typewriters, cash registers, and other electrically operated office equipment	40-60 g's
Moderately Delicate	
Television receivers, aircraft accessories	60-85 g's
Moderately Rugged	
Laundry equipment, refrigerators, appliances	85-115 g's
Rugged	
Machinery	115 g's and up

Figure 7.7 shows a plot of the numerical value of the amplification factor, A_m, for the half sine shock pulse contained in Figure 6.11. Figure 7.8 shows the amplification factors associated with other pulse shapes. An inspection of this figure shows that the rectangular pulse produces the greatest shock amplification factor. If such a pulse could be generated we could use it to produce a "worst case" shock testing procedure utilizing the relationship between G_m, τ, and Δv as shown in Figure 7.9a. Unfortunately, it is almost impossible to mechanically create a shock pulse with an instantaneous "rise time," but we can create a trapezoidal shock pulse with a rapid rise time, which is nearly as good[3] as the rectangular pulse. The trapezoidal pulse is shown in Figure 7.9b.

Based on the approximation of the trapezoidal pulse to the relationship of Equation 7.1, a fragility testing method called the **Damage Boundary Curve** has been developed(11). This is essentially a two dimensional index of fragility which takes both the amplitude and duration of the shock into account. A sample Damage Boundary Curve is shown in Figure 7.10.

[2]If we had a triangular shock pulse, the shape factor would be 1/2 (i.e., Area = 1/2 base x height).

[3]The amplification factor for the trapezoidal pulse is numerically equal to that of the rectangular pulse, except at large f_1/f_2 values where it is slightly lower.

*Courtesy of Dow Chemical Co.

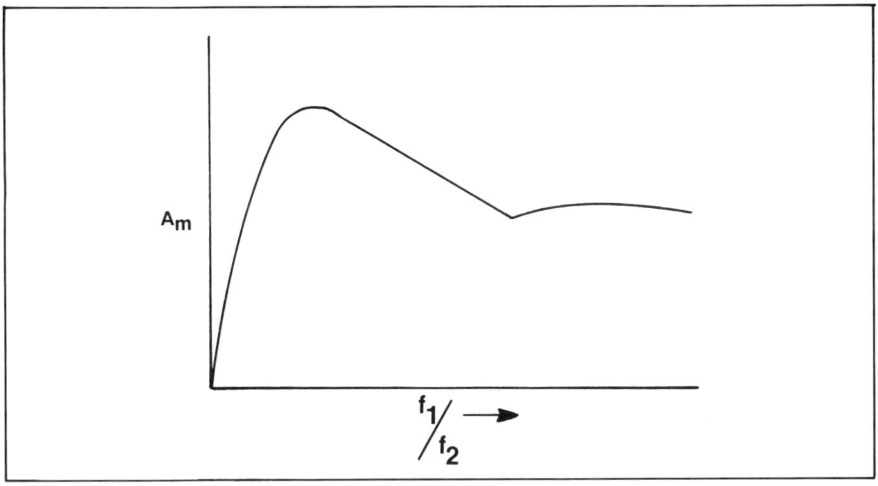

Figure 7.7. Amplification for the Half Sine Shock Pulse. (11)

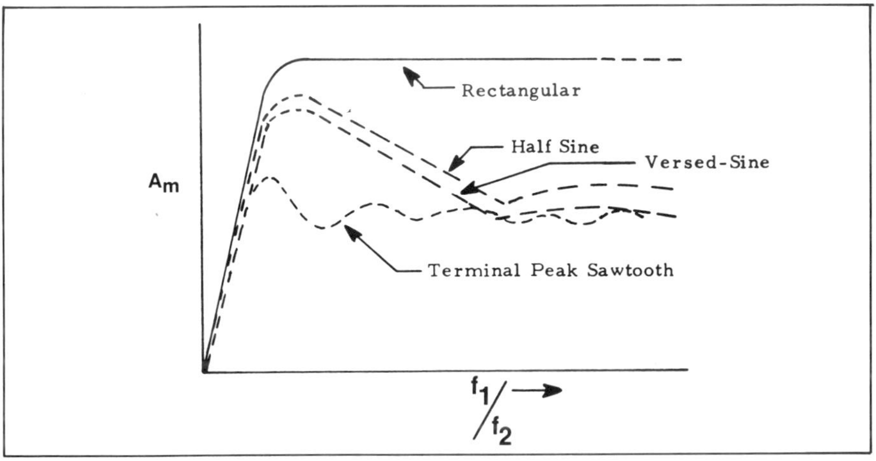

Figure 7.8. Amplification Curves for Various Shock Pulse Shapes. (11)

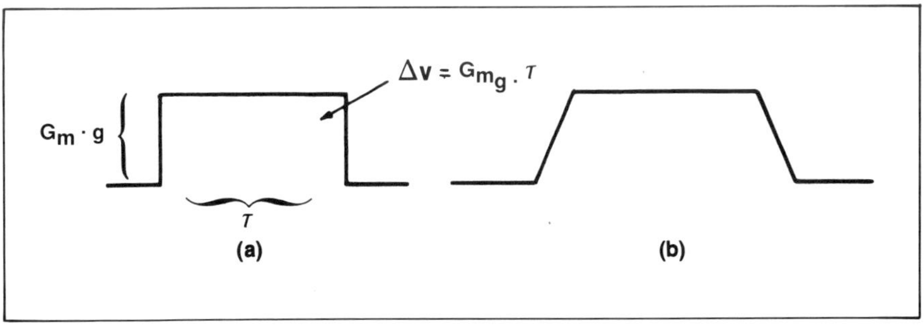

Figure 7.9. Rectangular and Trapezoidal Shock Pulses.

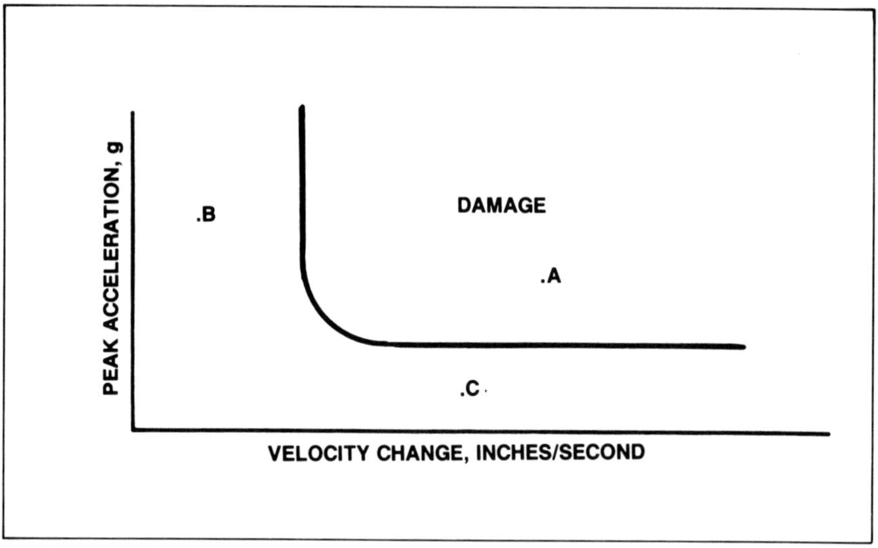

**Figure 7.10. A Damage Boundary Curve.
Point A; Damage,
Point B; No Damage,
Point C; No Damage.**

The Damage Boundary Curve relates peak shock acceleration on the vertical axis and total velocity change on the horizontal axis to product damage. A point or combination of peak acceleration and velocity change that falls above and to the right of the curve results in product damage. Point A is such a combination. However, Points B and C will not result in damage. Point B, although it represents a relatively high acceleration level, also represents a very low velocity change. From the relationship shown in Figure 7.9a, a high acceleration level and a low velocity change can only result from a shock pulse of very short duration. The product and its components characterized by the DBC of Figure 7.10 would not be damaged by such a short pulse; they simply did not respond to such a rapid event. For Point C, although the shock pulse may have produced a response from the product or its components, the acceleration level was not great enough to cause damage.

An examination of the procedure for developing the Damage Boundary Curve for a product will assist our understanding of the features of this two dimensional fragility display.

7.2 Constructing a Damage Boundary Curve

The development of a damage boundary curve for a product is presented in full in ASTM D 3332-77, "Standard Test Methods for Mechanical-Shock Fragility of Products, Using Shock Machines"(21). We will outline the procedure here so as to highlight some of the important features of the DBC.

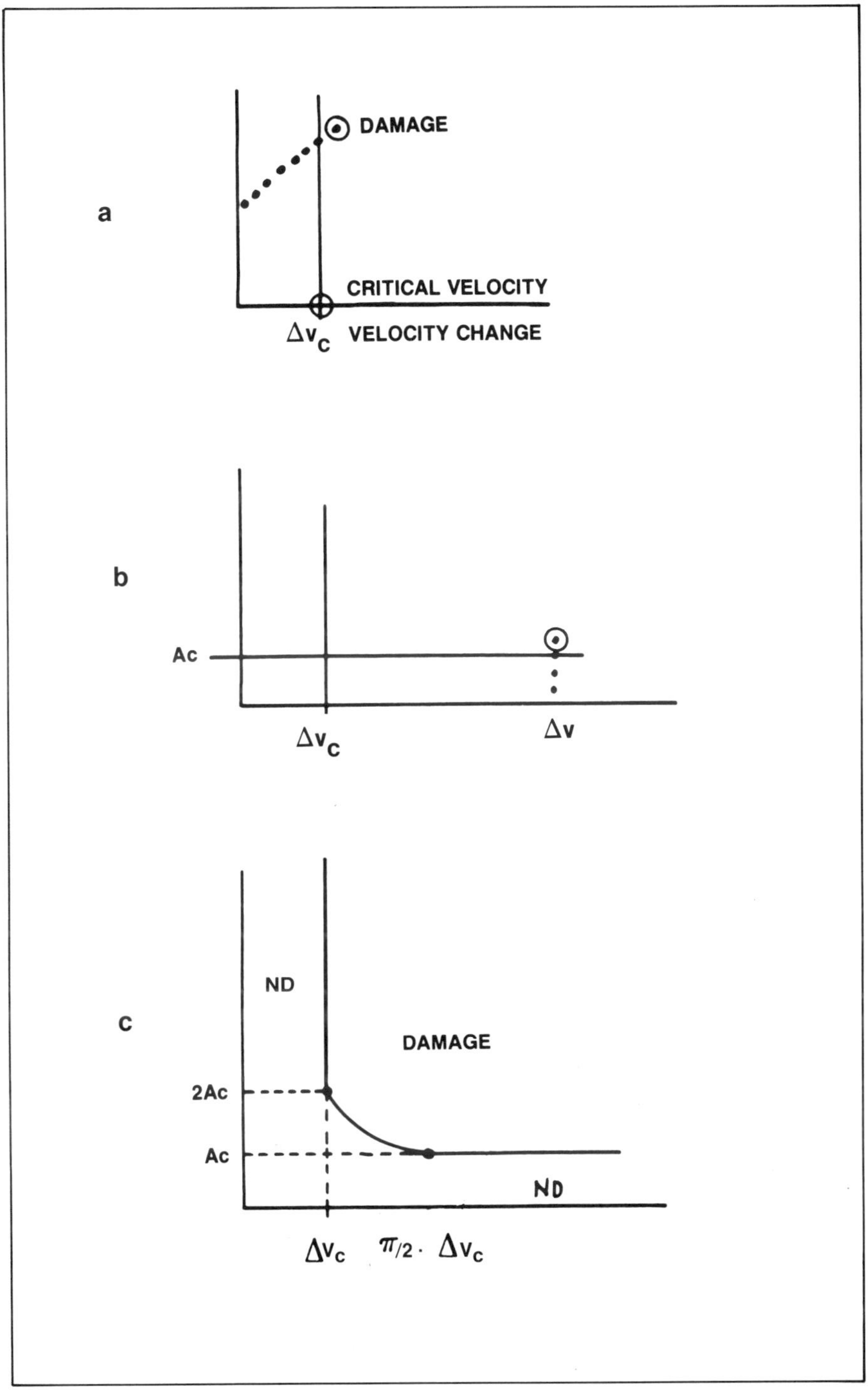

Figure 7.11. Construction of a Damage Boundary Curve.

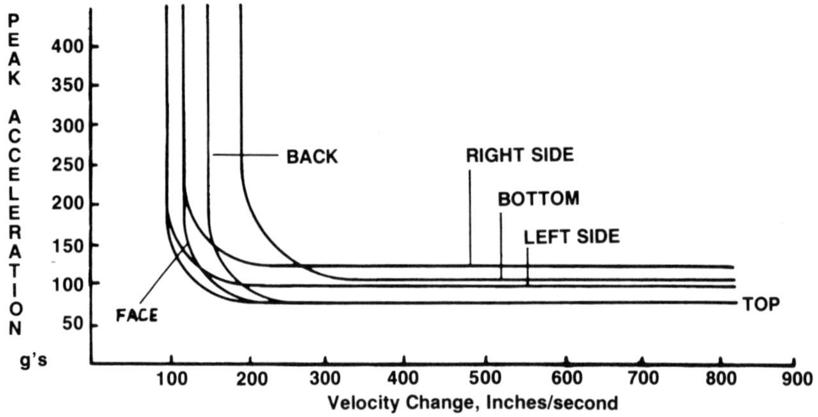

Figure 7.12. Damage Boundary Curves Representing Various Product Orientations. (14)

The first step in developing a DBC is to determine the **critical velocity change** (Δv_c) for a product. To do this, a shock machine programmed to produce shock pulses of very short duration (usually 2 milliseconds) is used. The product is affixed to the table of the shock machine and dropped onto the programmed impact surface from a series of drop heights, each slightly larger than the preceding, until the product is damaged. Each drop height corresponds to a point (Δv, G_m) on the curve shown in Figure 7.11a. The point where product damage occurs is circled. The critical velocity change, Δv_c, is the velocity change of the point immediately preceding the damage point. A vertical line is drawn through Δv_c.

The second step is to reprogram the shock machine so that it will produce low acceleration, trapezoidal shock pulses. This is usually accomplished using a gas piston as the impact surface. High gas pressure within the piston produces a short trapezoidal pulse with a relatively high acceleration level; low gas pressures produce longer duration pulses at lower acceleration levels.

The shock machine must also be set to drop onto the gas piston from a constant drop height. This drop height is chosen such that the velocity change exceeds $\frac{\pi}{2} \cdot \Delta v_c$ (this is done so that we will be operating on the flat, right hand part of the DBC). Having chosen an appropriate drop height, an undamaged product is again affixed to the table of the shock machine and a series of drops run. Between each drop the pressure in the gas piston is increased so that the acceleration level is increased while the velocity change remains constant (the pulse duration decreases with each succeeding drop). The Δv_c, G_m points are again plotted as shown in Figure 7.11b, with the damage point circled as before. The acceleration level of the point immediately preceding the damage point is the **critical acceleration,** A_c, and a horizontal line is drawn through that value.

The last step in drawing the DBC is to round off the lower left hand corner. A curve (part of an ellipse will do) is drawn between the points (Δv_c, 2 A_c) and ($\frac{\pi}{2}\Delta v_c$, A_c) as shown in Figure 7.11c. The resultant DBC for our product is shown in Figure 7.11d.

In the three step process, we have determined the DBC for just one orientation of our product. Since most products have different shock sensitivities in different orientations, several DBC's must be superimposed, as was done in Figure 7.12, to represent the overall fragility of the product.

For a practical example, the DBC test data given in Table 7.3 below has been translated into the curve shown in Figure 7.13, using the shock machine calibration values contained in Table 7.4. Notice that the shock machine was set to produce a Δv of 200 in/sec corresponding to a drop height of 24 inches during the second step, the critical acceleration determination.

TABLE 7.3

Test for Δv_c			Test for A_c		
Drop No.	**Height**	**Result**	**Drop No.**	**Gas Pressure**	**Result**
1	2"	no damage	7	50 psi	no damage
2	3"	no damage	8	100 psi	no damage
3	4"	no damage	9	150 psi	no damage
4	5"	no damage	10	200 psi	no damage
5	6"	no damage	11	250 psi	no damage
6	7"	damage	12	300 psi	no damage
			13	350 psi	damage

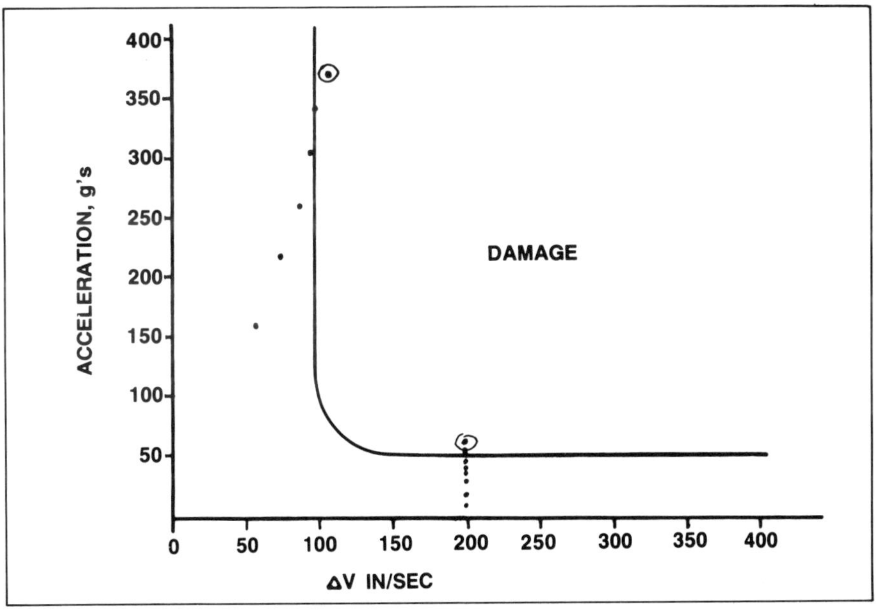

Figure 7.13. A Damage Boundary Curve Based on the Information of Tables 7.3 and 7.4.

TABLE 7.4
SHOCK MACHINE CALIBRATION VALUES(2)

2 ms Half-Sine Programmers (Bare Table)

Drop Height	ΔV	g's
2"	55 in/sec	160
3"	72 in/sec	215
4"	83 in/sec	260
5"	91 in/sec	305
6"	97 in/sec	340
7"	107 in/sec	370
8"	114 in/sec	400
9"	121 in/sec	430
10"	128 in/sec	455
11"	135 in/sec	480
12"	141 in/sec	515
13"	147 in/sec	540
14"	153 in/sec	560
15"	159 in/sec	580
16"	164 in/sec	605
17"	169 in/sec	620
18"	174 in/sec	
19"	179 in/sec	
20"	184 in/sec	

Gas Programmer (Bare Table)

Drop Height	ΔV	Gas Pressure	g's
3"	80 in/sec	50 psi	9
6"	110 in/sec	100 psi	18
12"	140 in/sec	150 psi	27
18"	170 in/sec	200 psi	36
24"	200 in/sec	250 psi	45
30"	230 in/sec	300 psi	54
36"	260 in/sec	350 psi	63
42"	290 in/sec	400 psi	72
		450 psi	81
		500 psi	90
		550 psi	99
		600 psi	108
		650 psi	117
		700 psi	126
		750 psi	135

7.3 Cushion Design and Product Protection

Information on the shock protection performance of commercially available cushioning material is usually presented in the form of a **cushion curve** like the one shown in Figure 7.15. Here, the maximum acceleration (or deceleration) at a cushioned product (on the vertical scale) is shown as a function of cushion thickness and static loading (horizontal scale). The static loading is defined as:

$$\text{Static loading} = \frac{\text{product weight}}{\text{bearing area of the cushion}}$$

Such curves are available for a variety of drop heights and usually for single and multiple impacts. The latter is necessary because many cushioning materials are permanently deformed by an impact and consequently their cushioning properties are reduced by each impact.

The information on product fragility contained in the Damage Boundary Curve may be mated with that of the cushion curve to produce a simple cushion design for protecting a product in distribution. An elementary example will illustrate how this may be done.

Suppose we have a cubic product weighing 50 pounds and measuring twelve inches on a side. The product's "fragility" is described by the damage boundary curve shown in Figure 7.14. In distribution we expect the product-package system to be dropped from a maximum height of 30 inches. We also expect that many of the packages will be dropped more than once from that maximum drop height. We want to design an Ethafoam™* 200 polyethylene cushion to protect our product.

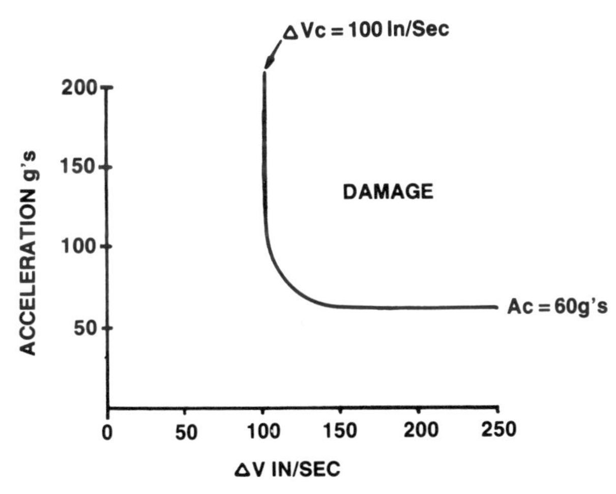

Figure 7.14. A Sample DBC.

*Ethafoam™ is a trademark of the Dow Chemical Company.

We will assume that the product is equally sensitive to damage on each of its six faces and this fragility is represented by the DBC of Figure 7.14. The product's critical velocity change is 100 in/sec and its critical acceleration is 60 g's.

The first step is to check if the product really needs cushioning. To do this we calculate the maximum velocity change possible in a 30 inch drop and compare this Δv against Δv_c.

$$\Delta v\ (30\text{ in.}) = 2\sqrt{2gh} = 2\sqrt{2 \cdot 386.4\text{ in/sec}^2 \cdot 30\text{ in.}} = 304\text{ in/sec}$$

$$304\text{ in/sec} > \Delta v_c = 100\text{ in/sec}$$

Since the velocity change in a 30 inch drop will[4] exceed the critical velocity change, we must cushion the product. If the expected velocity change was not larger than Δv_c, however, we would not have had to add cushioning.

The next step is to select the proper thickness for the Ethafoam™ 220 cushion. Using the cushion curve shown in Figure 7.15 (for 30 inch drops, 2-5 impacts) and the critical acceleration level of 60 g's, it is seen that 2 inch polyethylene, if loaded at 1.1 psi (pounds per square inch), will produce a maximum acceleration at the product of 60 g's when dropped 30 inches. This means that enough cushion must be present under the 12 x 12 inch side so that the static loading is 1.1 lb/in^2, i.e.

$$1.1\text{ lb/in}^2 = \frac{50\text{ lb.}}{\text{cushion area}}$$

The cushion area is, therefore,

$$\frac{50\text{ lb.}}{1.1\text{ lb/in}^2} = 45.5\text{ in}^2$$

Each face of the product should be cushioned with a 2 inch thick pad with a surface area of 45.5 square inches. A 2 in. x 6.74 in. x 6.74 in. would be appropriate[5].

Although the polyethylene cushioning material is non-linear we could estimate the maximum deflection and the shock duration using Equations 6.17, 6.19 and 6.24.

$$d_m = \frac{2h}{G_m} = \frac{2.30\text{ in}}{60} = 1.0\text{ inch}$$

$$k_2 = \frac{W_2\, G_m^2}{2h} = \frac{(50\text{ lb})\,(60)^2}{2 \cdot 30\text{ in.}} = 3000\text{ lb/in.}$$

---

[4]The minimum velocity change, $\sqrt{2gh}$, is 152 in/sec which is also larger than Δv_c, so it is 100% probable that $\Delta v > \Delta v_c$ in a 30 inch drop.

[5]Many use the "buckling" criterion to test the cushion's thickness versus area. To avoid a "tall, skinny" cushion which would not compress uniformly, the area should be larger than (1.33 x thickness)^2.

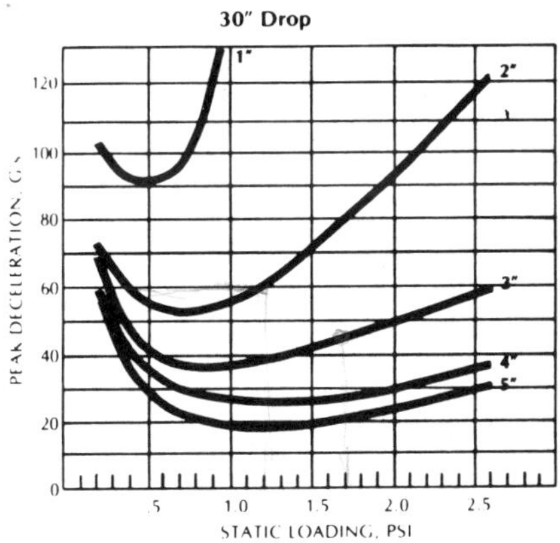

Figure 7.15. Cushion Curves for Dow Ethafoam™ 220.

$$\tau = \frac{\pi}{\sqrt{\frac{k2g}{W_2}}} = \frac{\pi}{\sqrt{\frac{300 \text{ lb/in} \cdot 386.4 \text{ in/sec}^2}{50 \text{ lb.}}}} = 0.02 \text{ second or } 20 \text{ ms}$$

We may also investigate the implications of this simple cushion design on the package weight, volume and surface area. The total weight of the six cushions may be computed since the density of Ethafoam™ 220 is known to be 2.2 pounds/cubic foot:

$$(6 \text{ cushions}) \times \frac{(2 \text{ in. x } 45.5 \text{ in.}^2)}{1728 \text{ in}^3/\text{ft}^3} = 0.316 \text{ ft}^3$$

$$(0.316 \text{ ft}^3) \times (2.2 \text{ lb/ft}^3) = 0.695 \text{ lbs.}$$

Two inch thick cushions are attached to each of the six faces of the product. The inside dimensions of the box needed to contain the product and cushions are 16 in. x 16 in. x 16 in. (two 2-inch cushions added to each dimension). Thus, each face of the box is 16 in. x 16 in. = 256 square inches. The minimum area to be covered by corrugated board is:

$$(6 \text{ sides}) \times (256 \text{ in}^2/\text{side}) = 1536 \text{ square inches}$$

The volume of this box is:

16 in. x 16 in. x 16 in. = 4096 cubic inches (2.37 cubic feet)

The density of the package system is (ignoring the corrugated weight):

$$\frac{50 \text{ lb } + 0.695 \text{ lb.}}{4096 \text{ in}^3} = 0.0124 \text{ lb/in}^3 \text{ (or } 21.4 \text{ lb/ft}^3\text{)}$$

Suppose in the example we had chosen three-inch Ethafoam™ 220 instead of two inch. For 3-inch foam the static loading is 2.6 psi, so our working area becomes:

$$\frac{50 \text{ lb.}}{2.6 \text{ psi}} = 19.2 \text{ in}^2 \text{ (greater than the 15.9 in}^2 \text{ for the buckling criterion)}$$

The total weight of the six 3-inch cushions is:

$$(6 \text{ cushions}) \times \left(\frac{3 \text{ in} \times 19.2 \text{ in}^2}{1728 \text{ in}^3/\text{ft}^3}\right) \times (2.2 \text{ lb/ft}^2) = 0.440 \text{ lb.}$$

Three-inch thick cushions are attached to each of the six faces of the product. The inside dimensions of the box needed to contain the product and cushions are:

18 in. x 18 in. x 18 in. (two 3-inch cushions added to each dimension)

Thus, each face of the box is 18 in. x 18 in. = 324 square inches. The minimum area to be covered with corrugated is:

(6 sides) x (324 square inches) = 1944 square inches

The volume of the box is:

18 in. x 18 in. x 18 in. = 5832 cubic inches (3.38 cubic feet)

The density of the package system is:

$$\frac{50 \text{ lb} + 0.440 \text{ lb.}}{5832 \text{ in}^3} = 0.0086 \text{ lb/in}^3 \text{ (or 15.0 lb/ft}^3\text{)}$$

In summary, we see that by choosing three-inch cushioning instead of two inch:

1. The surface area of the box was increased from 1536 square inches to 1944 square inches, an *increase of 27%.*

2. The volume of the package was increased from 2.37 cubic feet to 3.38 cubic feet, an *increase of 43%.*

3. The density of the package was reduced from 21.4 lb/ft^3 to 15 lb/ft^3, a *decrease of 30%.*

All of these factors are reflected in the cost of packaging, shipping and handling, so it is obvious that the thinnér cushion is superior in terms of distribution economics.

7.4 Cushions and Vibrations

Another example will illustrate how vibration considerations may be factored into the design of package cushioning.

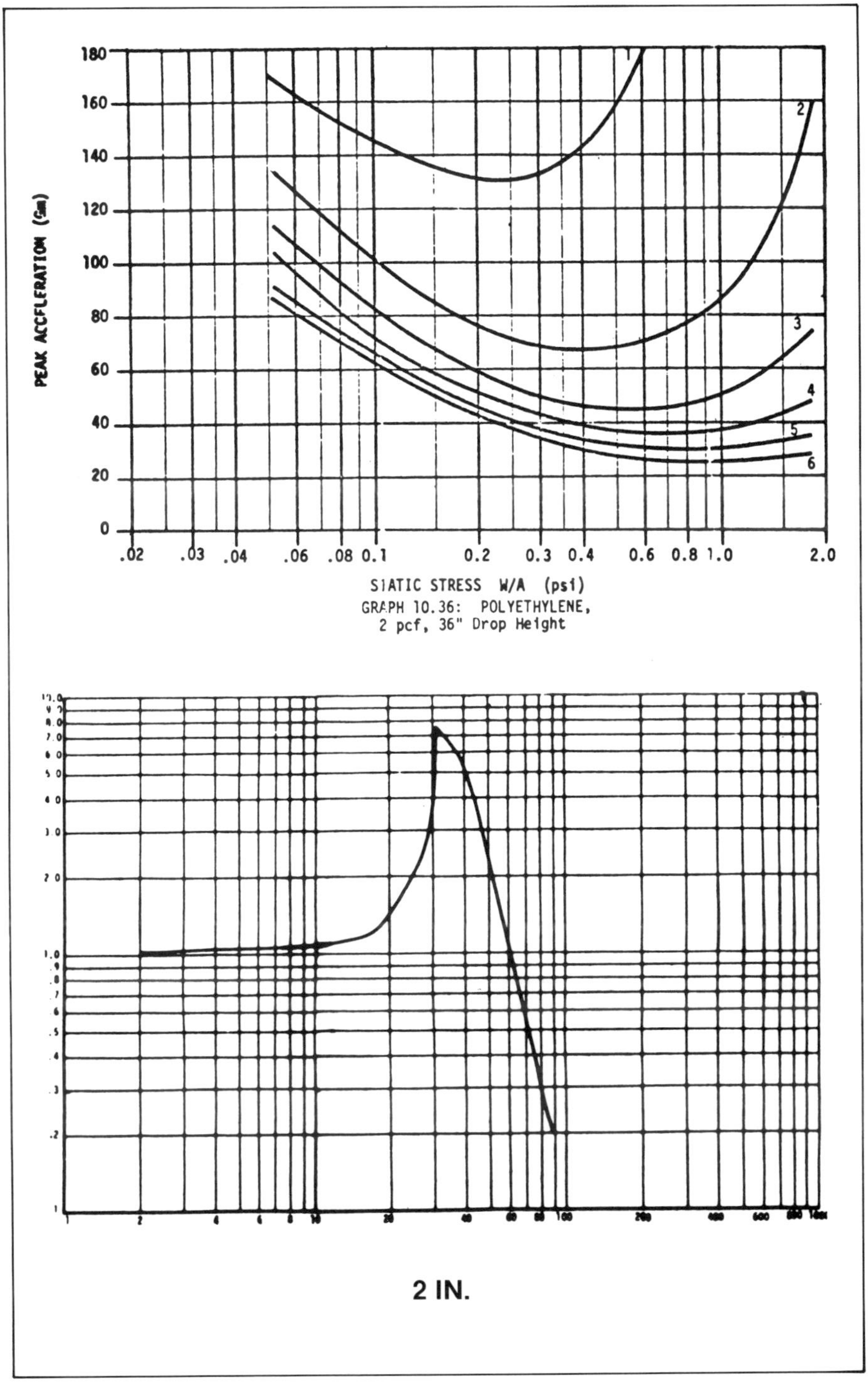

Figure 7.16. Polyethylene Cushion and Transmissibility Curves. (19)

Suppose we again have a cubic product measuring five inches on a side and weighing 24 pounds. The critical veocity for this item is 200 in/sec and the critical acceleration is 95 g's. We expect drops from a maximum height of 36 inches. The cushion curve we need is shown in Figure 7.16a (this is taken from the "Military Standardization Handbook, package Cushioning Design, MIL-HDBK-3043,"(19)). Examining the maximum velocity change expected in a 36 inch drop.

$$\Delta v = 2\sqrt{2 \cdot 386.4 \text{ in/sec}^2 \cdot 36 \text{ in.}} = 344 \text{ in/sec}$$

we see that we must cushion the product. From the curve in Figure 7.16a, the critical acceleration of 95 g's requires a 2-inch cushion with a static stress or loading of 1.2 psi. Our cushion area is then,

$$\text{Cushion Area} = \frac{24 \text{ lb.}}{1.2 \text{ lb/in}^2} = 20 \text{ in}^2$$

A 2 in. x 4 in. x 5 in. cushion will fit nicely on each of the product's faces. However, before considering this design complete we must examine the product-cushion system's vibration characteristics. Figure 7.16b shows the **transmissibility** (or **magnification)** of the 2-inch cushion loaded at 1.2 psi. Notice that the maximum transmissibility occurs at 30 Hz with values over 2.0 between 25 Hz and 45 Hz. Before proceeding further with the cushion design, the packaging technologist must examine the distribution environment through which this system will pass. If forcing frequencies between 25 and 45 Hz are present, the product-cushion system should be subjected to laboratory vibration testing to detect possible product damage.

PROBLEM SET, SECTION 7

1. Calculate the velocity change which occurred during each of the pulses shown below.

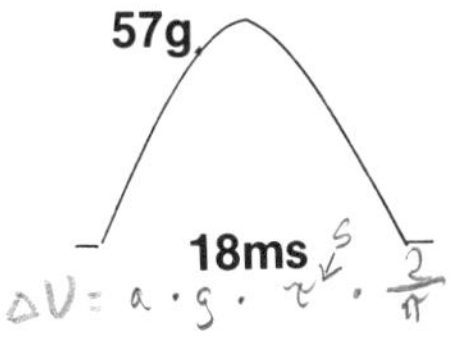

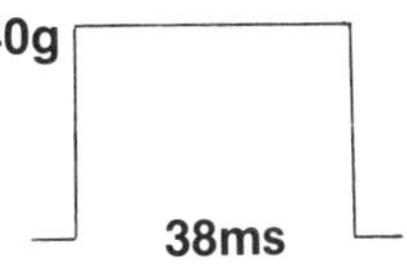

2. Plot a damage boundary using the following data, and the shock machine calibration values in Table 7.4.

TEST FOR CRITICAL VELOCITY CHANGE

Drop No.	Height	Results
1	2"	none
2	3"	none
3	4"	none
4	5"	none
5	6"	none
6	7"	none
7	8"	none
8	9"	damage occurred

TEST FOR CRITICAL ACCELERATION

Drop No.	Gas Pressure	Results
9	50 psi	none
10	100 psi	none
11	150 psi	none
12	200 psi	none
13	250 psi	none
14	300 psi	none
15	350 psi	none
16	400 psi	none
17	450 psi	damage occurred

3. Plot the damage boundary curve for a product from the data given below.

Test for Δv_c			Test for A_c		
Drop No.	**Height**	**Result**	**Drop No.**	**Gas Pressure**	**Result**
1	2"	no damage	12	50 psi	no damage
2	3"	no damage	13	100 psi	no damage
3	4"	no damage	14	150 psi	no damage
4	5"	no damage	15	200 psi	no damage
5	6"	no damage	16	250 psi	no damage
6	7"	no damage	17	300 psi	no damage
7	8"	no damage	18	350 psi	damage
8	9"	no damage			
9	10"	damage			

4. A 10-inch cubic product weighing 50 pounds has the damage boundary curve shown below. Design an Ethafoam™ cushion which will protect the product if it is expected to experience multiple drops from a maximum height of 30 inches in the distribution system. (Use the cushion curves following this problem set.)

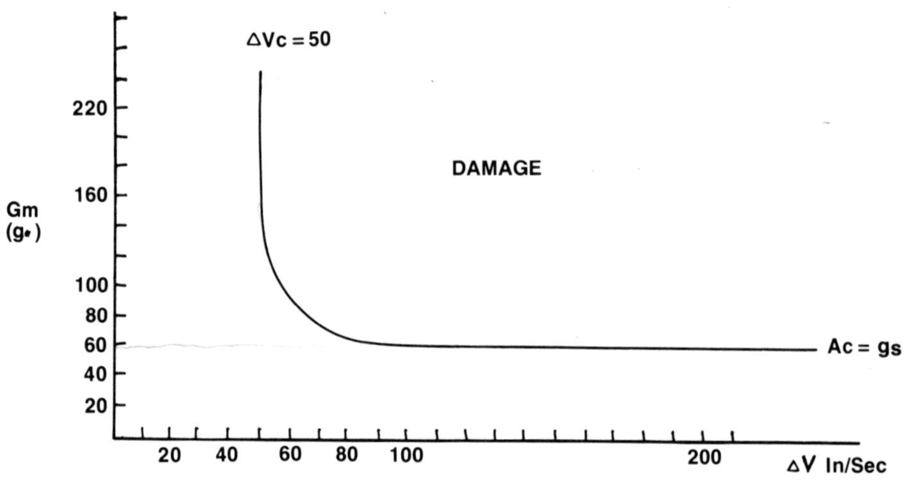

5. Design a 1.2 PCF Expanded Polystyrene (DYLITE) cushion to protect the product described in Problem 4. Assume the distribution conditions are the same.

6. An item weighing 200 lbs. has a critical acceleration of 20 g's ($A_c = 20$) and a critical velocity (Δv_c) of 100 in/sec. The item measures 9" x 11" x 17". Select an Ethafoam™, find the necessary thickness and determine the area needed for each face of the item if it is expected that the item will suffer multiple drops from a maximum height of 24 inches in the distribution environment.

7. The damage boundary curve for an 1100 lb. electrical generator is shown below. Select a foam cushion to protect the device if it will suffer multiple drops from heights not exceeding one foot in the distribution system. (The generator measures 2' x 3' x 6'.)

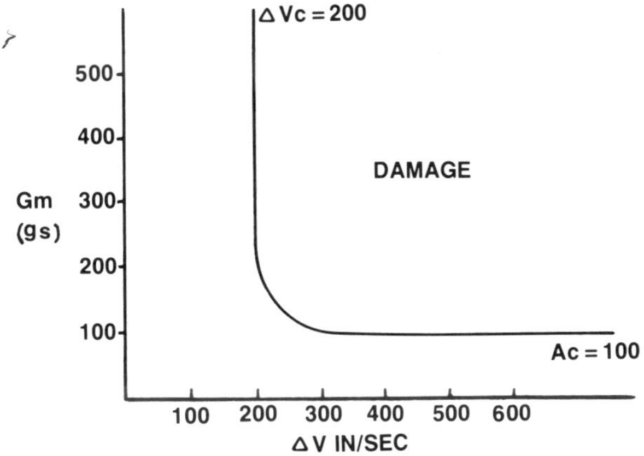

8. An article weighing 35 pounds mesures 6" x 10" x 12". It has a critical acceleration (fragility) of 45 (A_c = 45) and a critical velocity of 145 in/sec. It needs to be cushioned to protect it in a situation where it may be dropped on any face from 36 inches. Find the minimum thickness of Ethafoam™ 220 needed to cushion the item and compute the area to be covered with the cushioning material. Repeated impacts are expected.

9. For the cushion of Problem 8, estimate the spring constant k, the shock duration τ, and the cushion deflection d_m using free fall shock formulas.

10. A battery charger weighs 5.1 lbs. and measures 8-1/4" x 5-7/16" x 4-9/16". Its fragility is shown in the damage boundary curve below. Design Ethafoam™ cushioning which will protect the battery charger during multiple drops from a maximum height of 30 inches. The device has a critical element with a f_n = 35 Hz, which has been tested and found to survive 0.3 g at frequencies of 50-100 Hz. Check your foam cushioning for adequate vibration protection at the 70 Hz, 0.25 g forcing frequency found in the distribution environment.

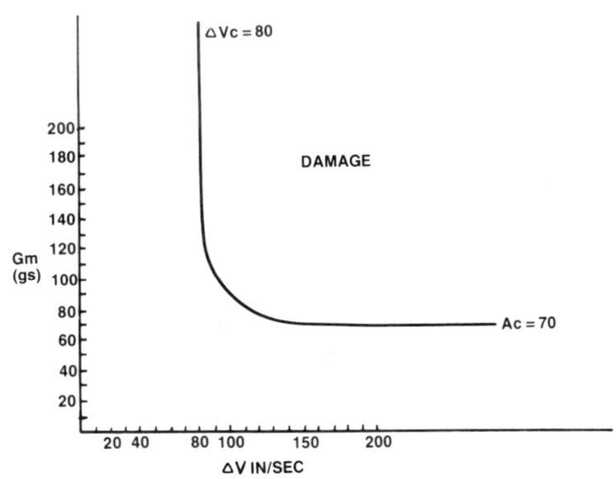

11. The product for which the damage boundary curve was constructed in Problem 3 weighs 50 lbs. and measures 9" x 8" x 9". Design a polystyrene cushion for this product; it is expected that the product will suffer multiple drops from a maximum height of 30 inches. What dimensions are each of the cushions? What are the minimum inside dimensions of the container which will hold the product and its cushions? What is the total weight of the product and its cushions?

12. A 35 lb. product measuring 4" x 6" x 12" can withstand, at maximum, a shock of 35 g's on any of its six faces. Compute the linear spring constant of the springs necessary to protect the product in a 36 inch fall. What will be the maximum deflection of the linear springs? If the working length of each spring is 78% of its total length, find the inner dimensions of the container needed to hold the item and the springs.

13. A product weighs 28.0 lbs. and measures 6" x 6" x 8". This product has been thoroughly tested for shock; its damage boundary curve is described by the parameters: ΔV_c (critical velocity) = 250 in/sec and A_c (critical acceleration) = 30,912 in/sec^2. If we expect the product to suffer multiple drops on any face from a maximum height of 36 inches, what are the minimum inside dimensions of the container needed to hold the product and its cushions? Use Ethafoam™ 220 polyethylene foam. (1 g = 386.4 in/sec^2) (Don't forget to check for buckling) What are the dimensions of each cushion?

14. A 10 lb. item measuring 6 inches on a side is cushioned on each face with a linear undamped spring. Each spring is four inches long, has a spring constant k = 500 lb/in and a working length equal to 38% of its total length.

    Is the spring length (4 in.) adequate to prevent "bottoming out" if the maximum drop height expected is 36 inches?

Cushion Package Design Data

Cushioning curves for ETHAFOAM 220 ™

Figure 1
12″ Drop, 1st Impact

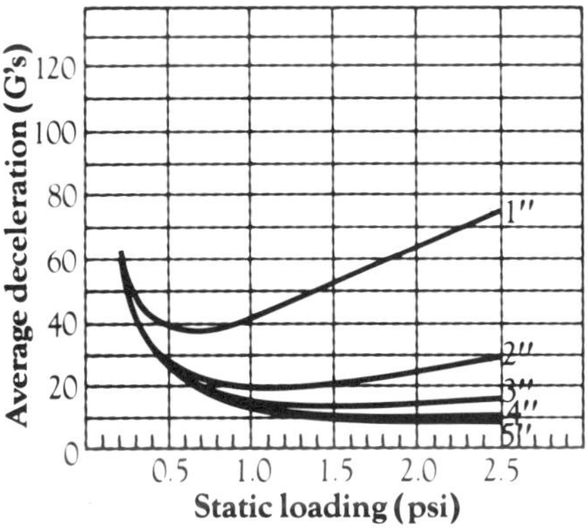

Figure 2
12″ Drop, 2-5 Impacts

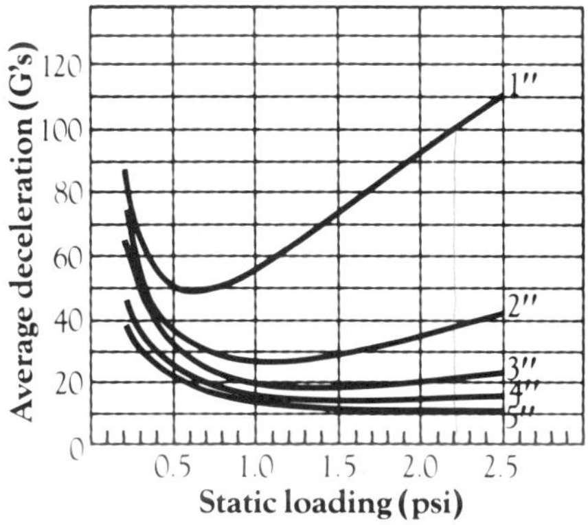

Figure 3
18″ Drop, 1st Impact

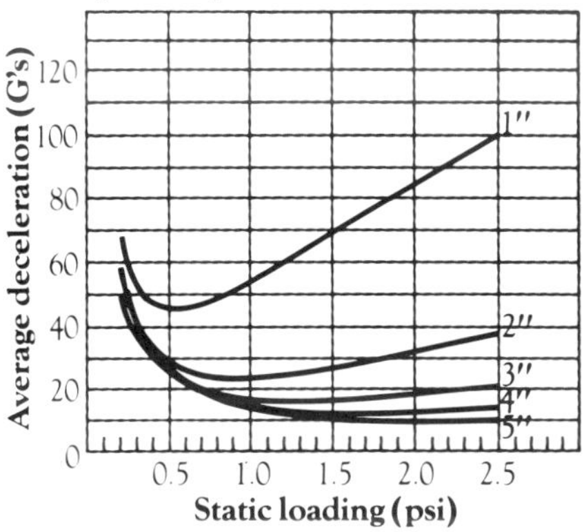

Figure 4
18″ Drop, 2-5 Impacts

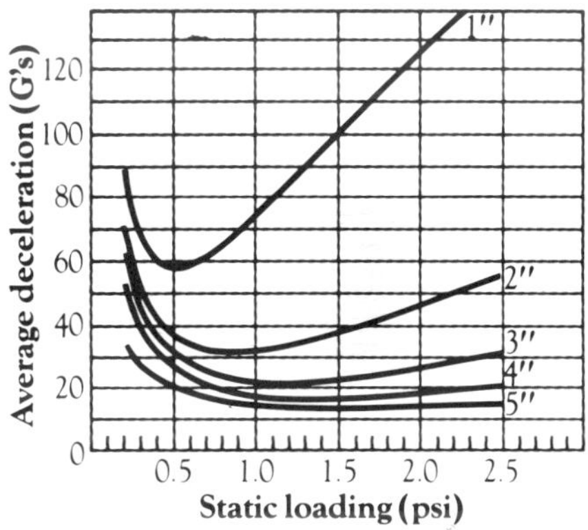

Figure 5
24″ Drop, 1st Impact

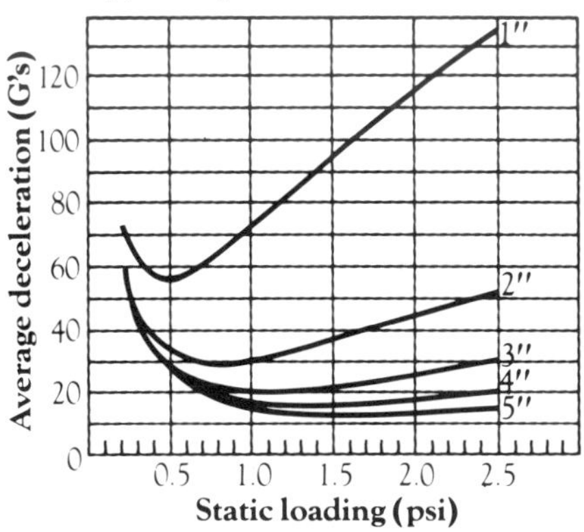

Figure 6
24″ Drop, 2-5 Impacts

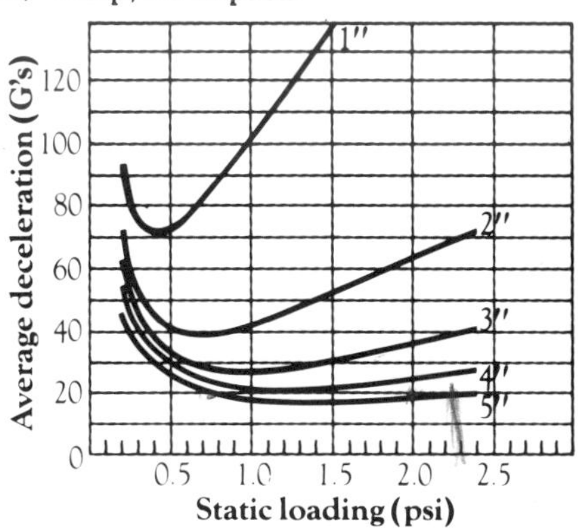

Courtesy of Dow Chemical Co.

Figure 7
30″ Drop, 1st Impact

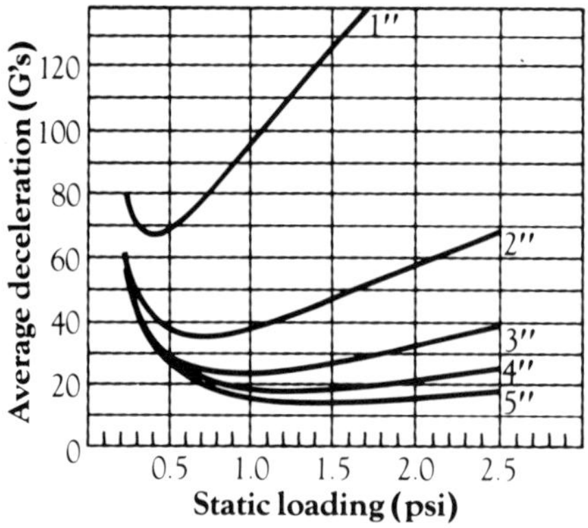

Figure 8
30″ Drop, 2-5 Impacts

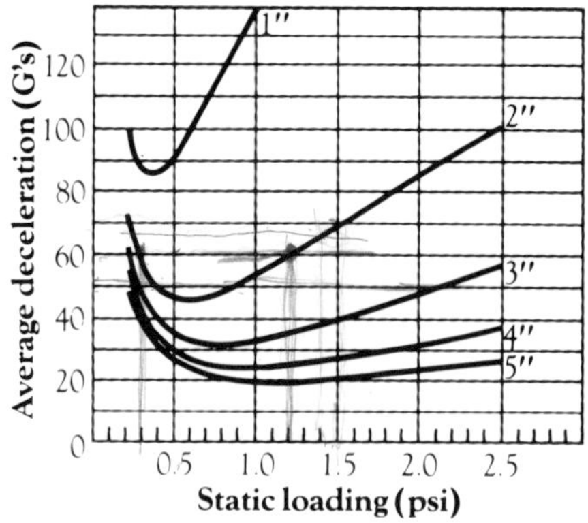

Figure 9
36″ Drop, 1st Impact

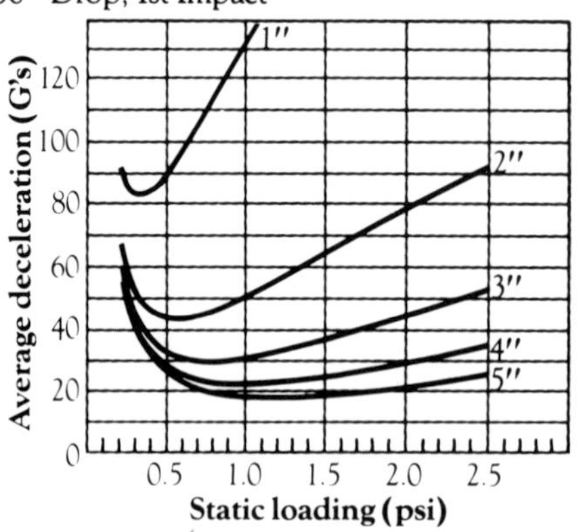

Figure 10
36″ Drop, 2-5 Impacts

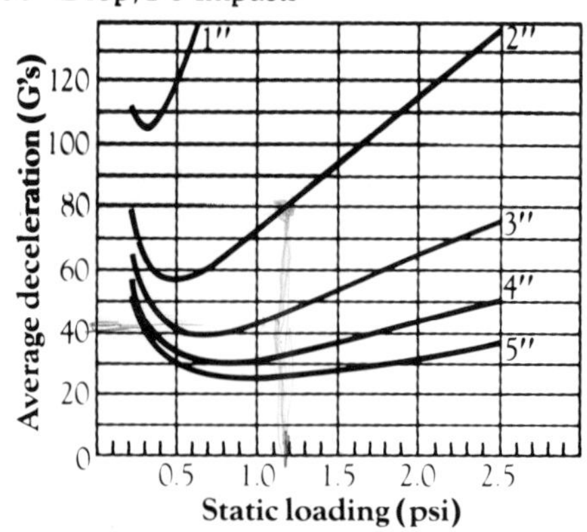

Figure 11
42″ Drop, 1st Impact

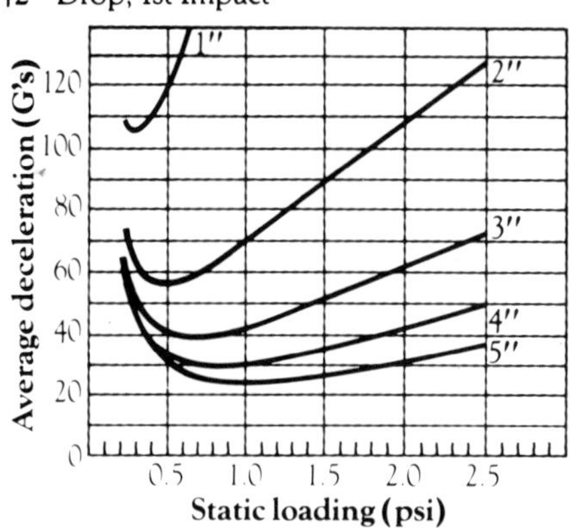

Figure 12
42″ Drop, 2-5 Impacts

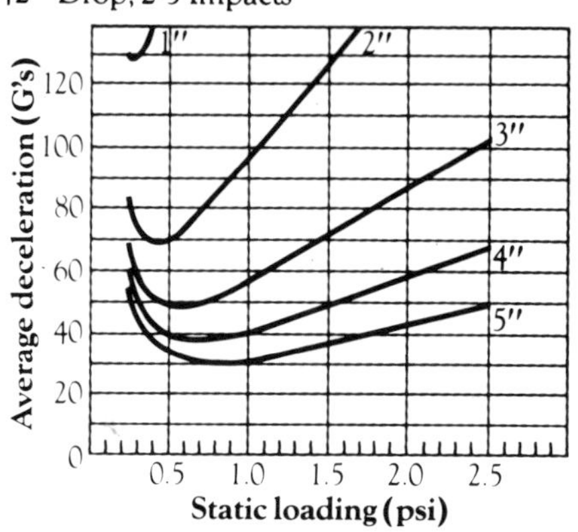

Figure 13
48″ Drop, 1st Impact

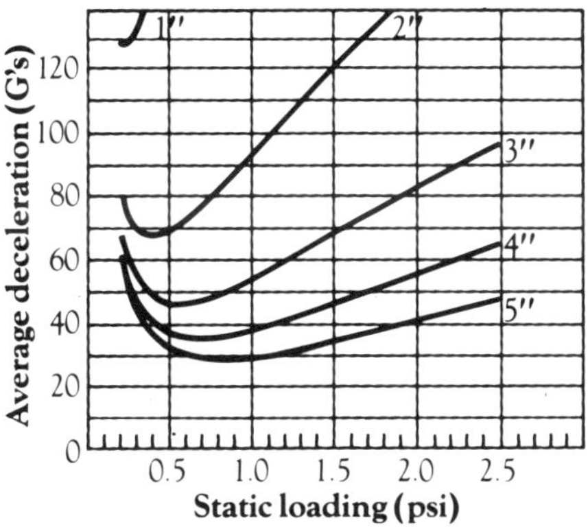

Figure 14
48″ Drop, 2-5 Impacts

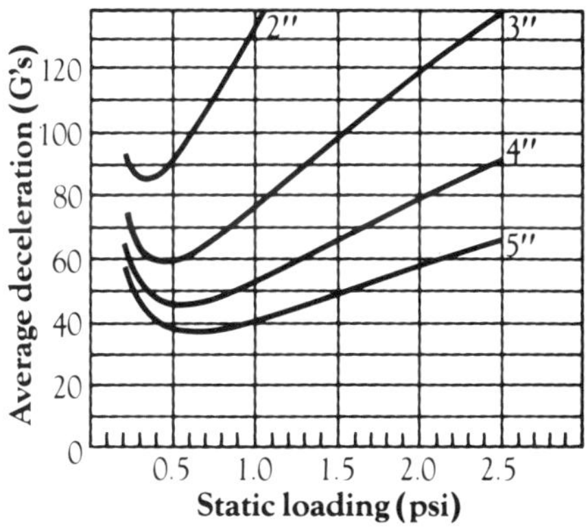

Cushioning curves for ETHAFOAM HS 45

Figure 15
12″ Drop, 1st Impact

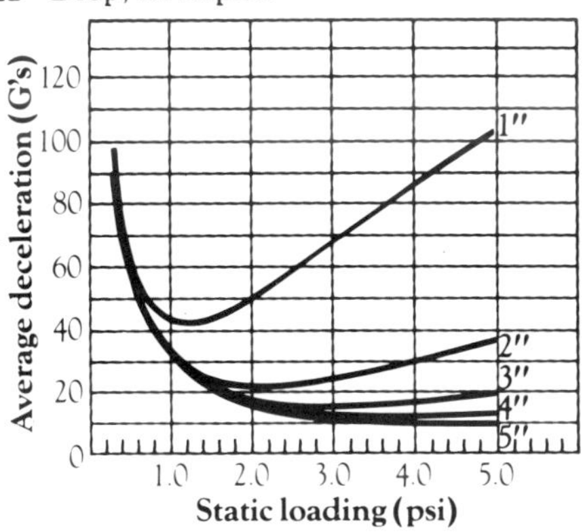

Figure 16
12″ Drop, 2-5 Impacts

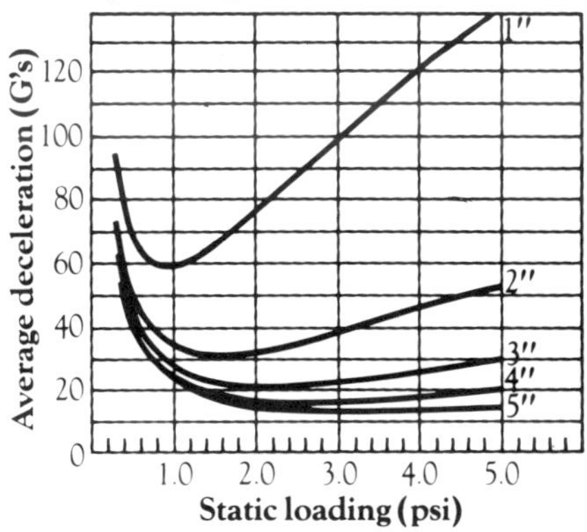

Figure 17
18″ Drop, 1st Impact

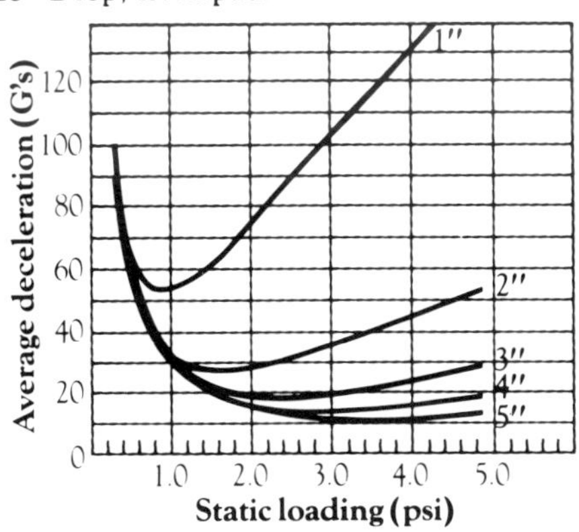

Figure 18
18″ Drop, 2-5 Impacts

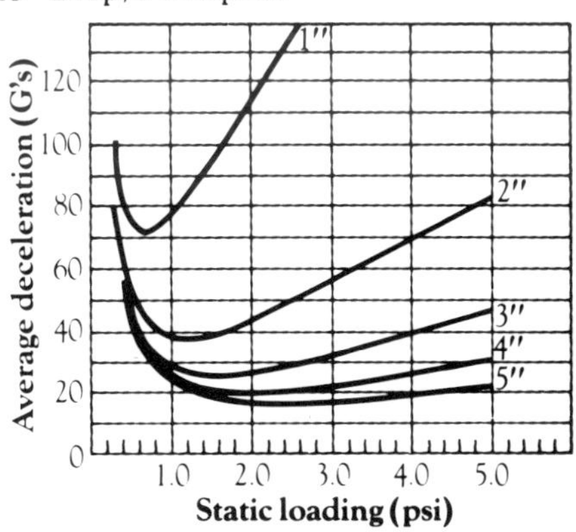

Figure 19
24″ Drop, 1st Impact

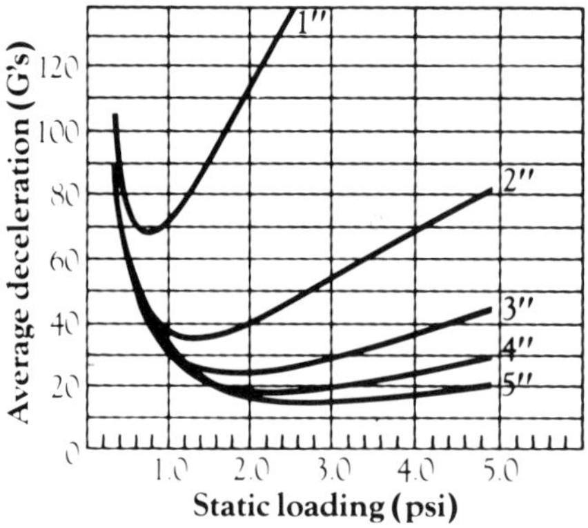

Figure 20
24″ Drop, 2-5 Impacts

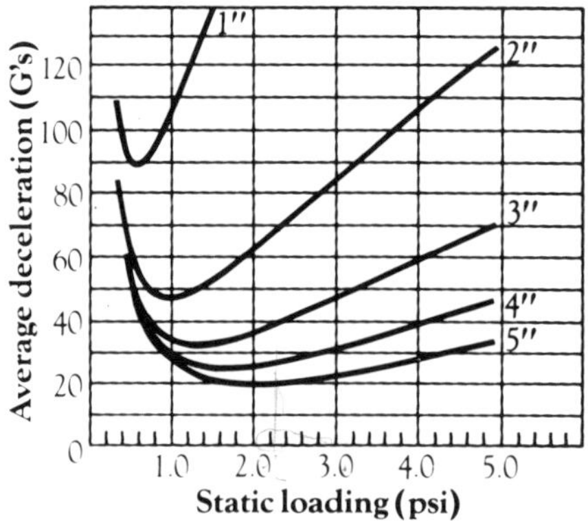

Figure 21
30″ Drop, 1st Impact

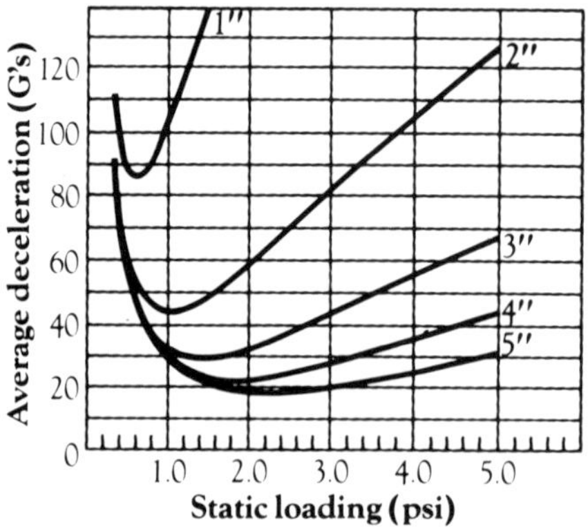

Figure 22
30″ Drop, 2-5 Impacts

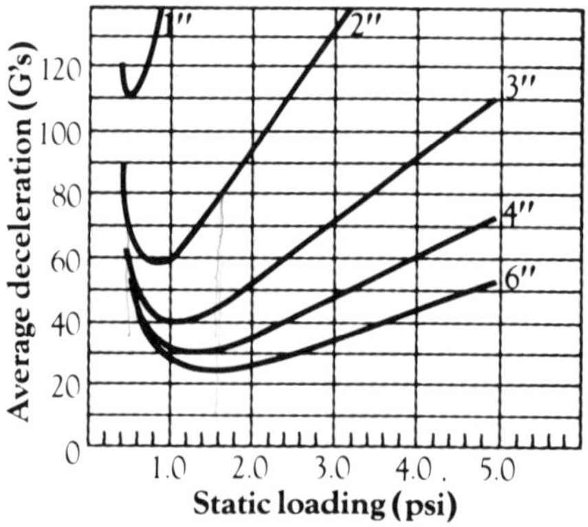

Figure 23
36″ Drop, 1st Impact

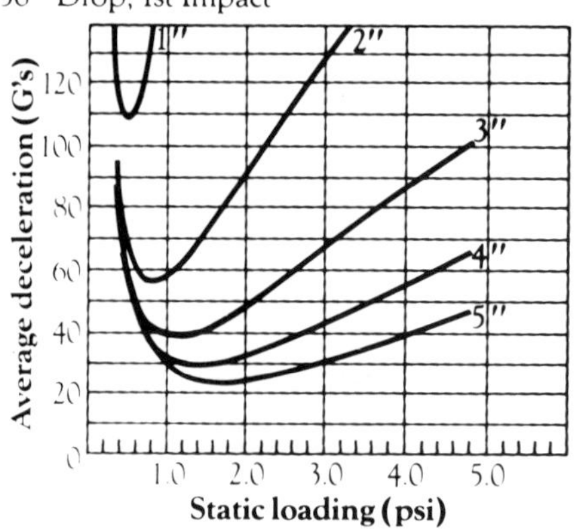

Figure 24
36″ Drop, 2-5 Impacts

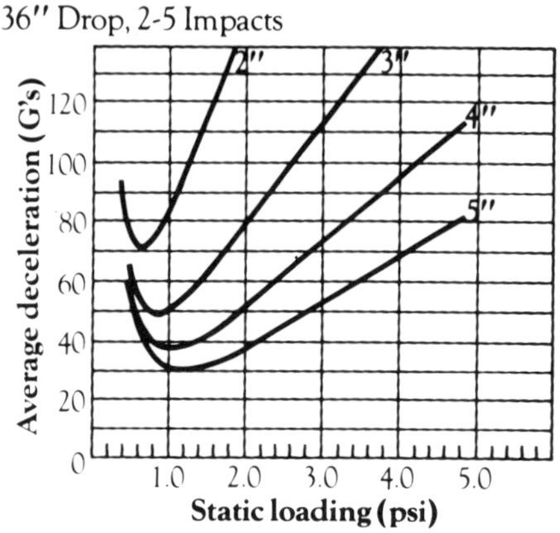

Cushioning curves for ETHAFOAM HS 70

Figure 25
12″ Drop, 1st Impact

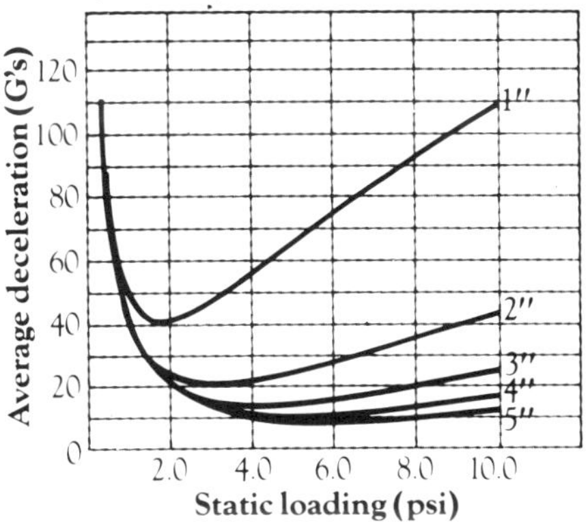

Figure 26
12″ Drop, 2-5 Impacts

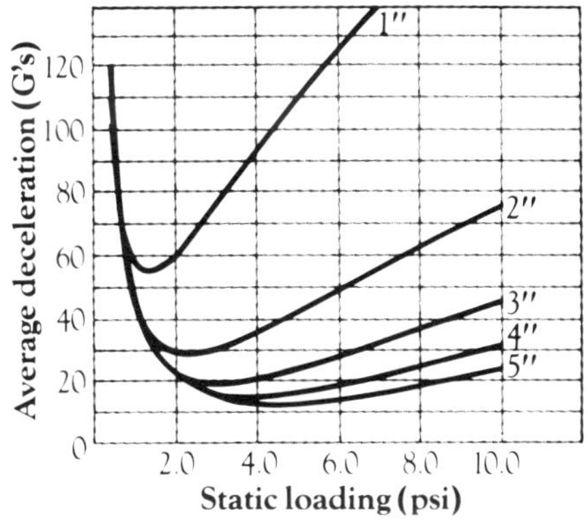

Figure 27
18″ Drop, 1st Impact

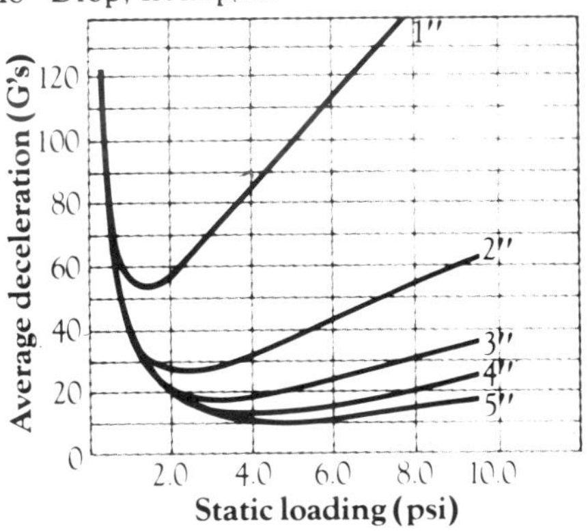

Figure 28
18″ Drop, 2-5 Impacts

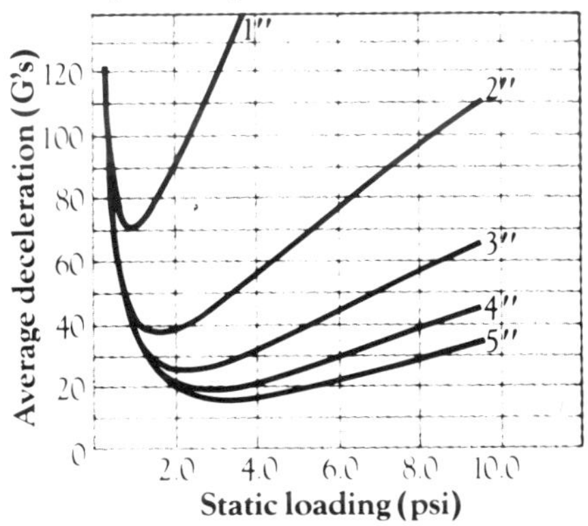

Figure 29
24″ Drop, 1st Impact

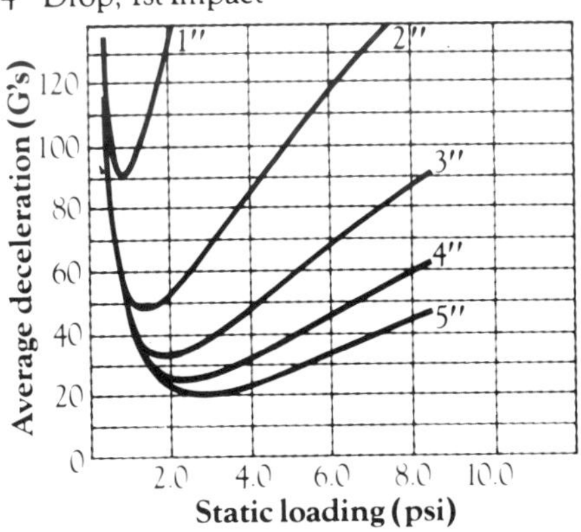

Figure 30
24″ Drop, 2-5 Impacts

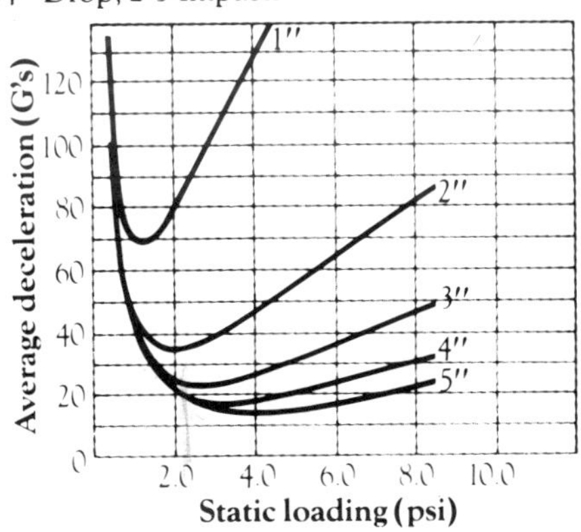

Figure 31
30″ Drop, 1st Impact

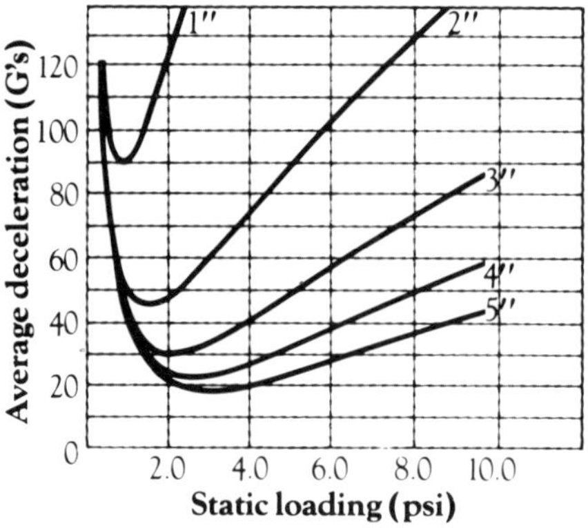

Figure 32
30″ Drop, 2-5 Impacts

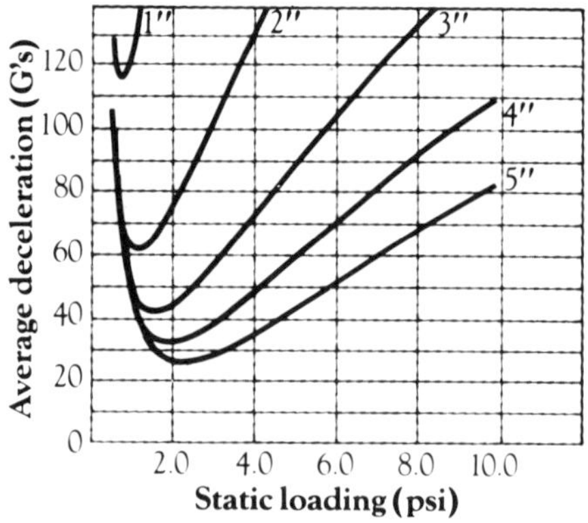

Figure 33
36″ Drop, 1st Impact

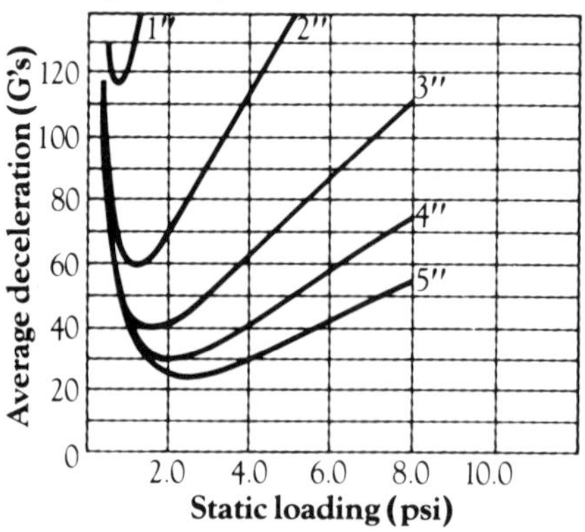

Figure 34
36″ Drop, 2-5 Impacts

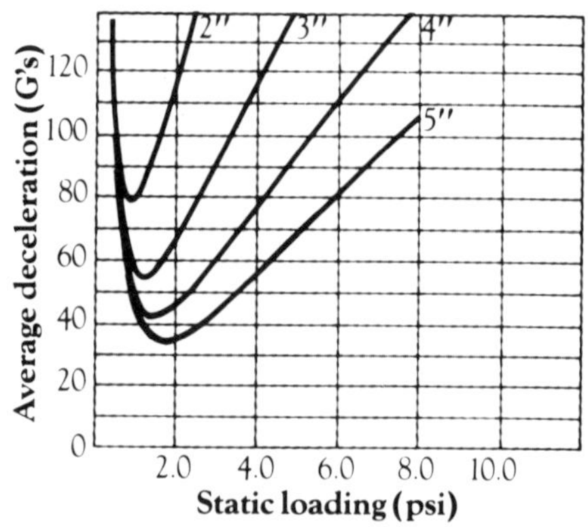

Cushioning curves for ETHAFOAM HS 120

Figure 35
12″ Drop, 1st Impact

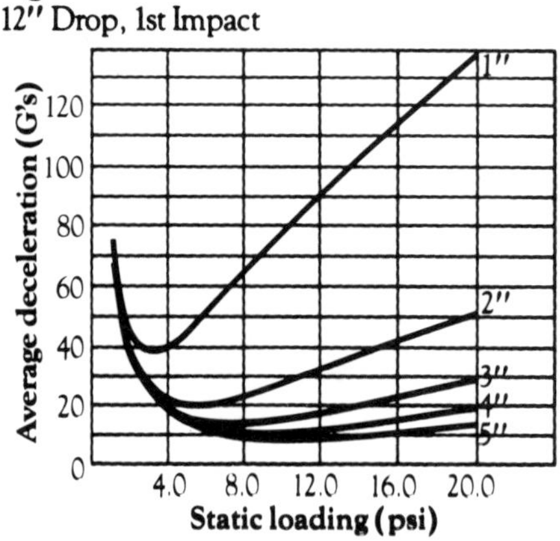

Figure 36
12″ Drop, 2-5 Impacts

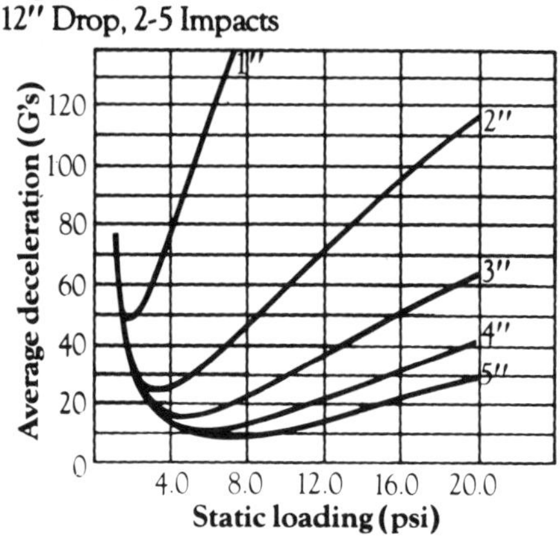

Figure 37
18″ Drop, 1st Impact

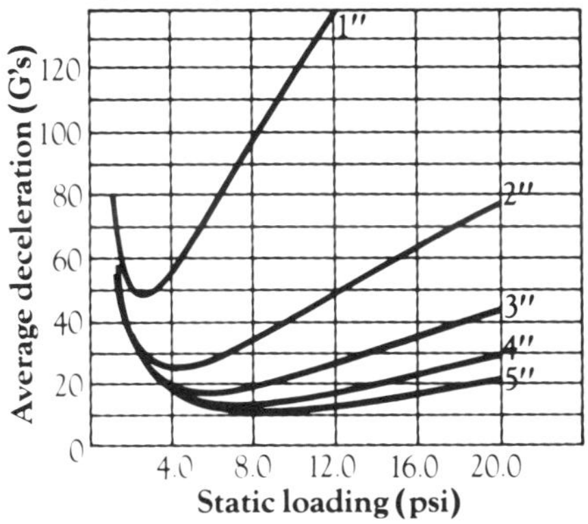

Figure 38
18″ Drop, 2-5 Impacts

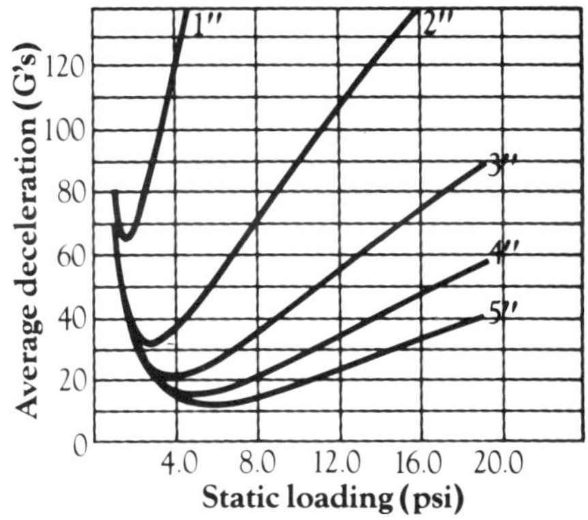

Figure 39
24″ Drop, 1st Impact

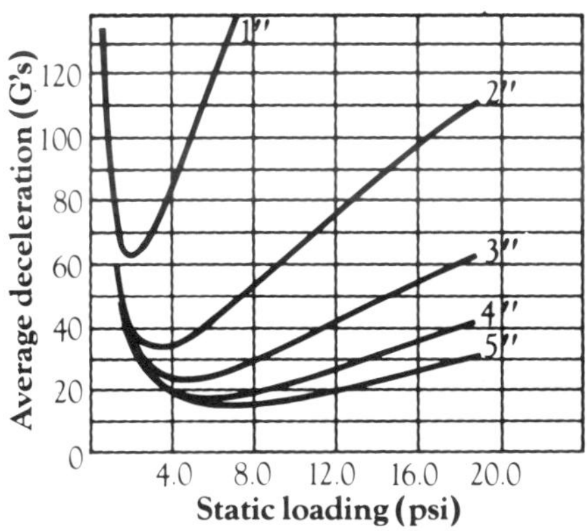

Figure 40
24″ Drop, 2-5 Impacts

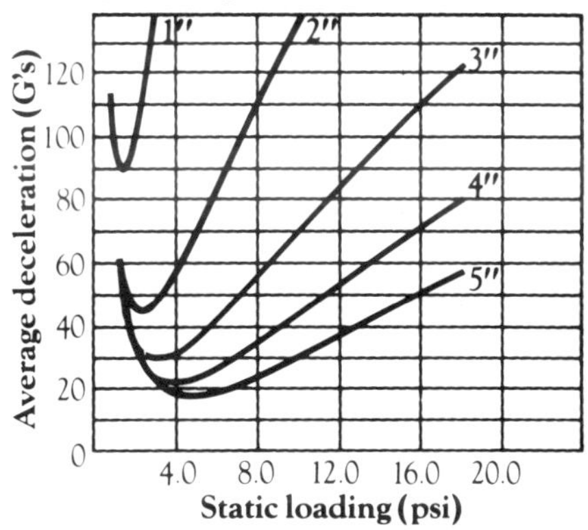

Figure 41
30″ Drop, 1st Impact

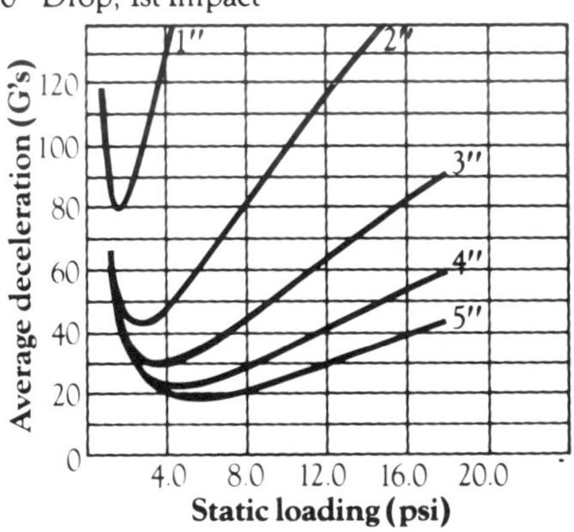

Figure 42
30″ Drop, 2-5 Impacts

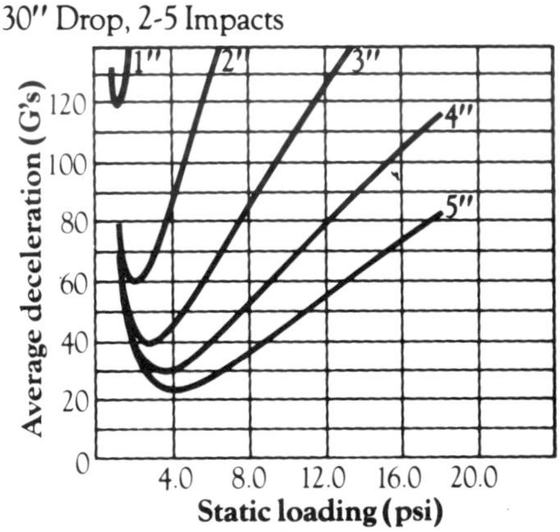

Figure 43
36" Drop, 1st Impact

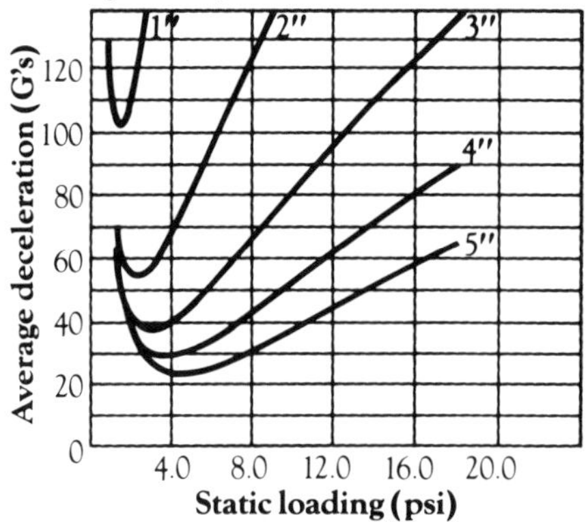

Figure 44
36" Drop, 2-5 Impacts

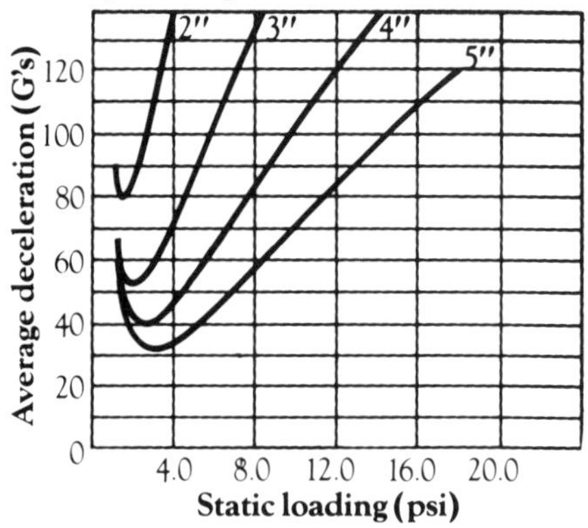

Amplification & Attenuation

FIGURE 61
Resonance*
ETHAFOAM 220

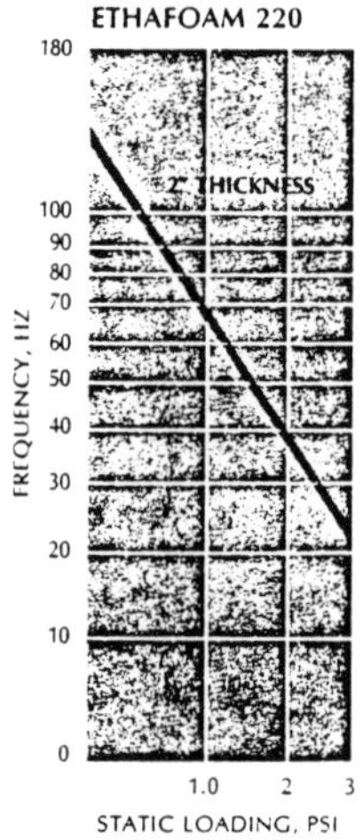

FIGURE 62
Resonance*
ETHAFOAM 400

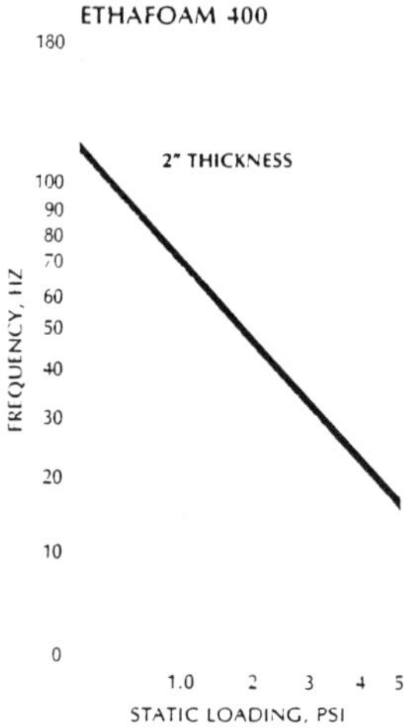

FIGURE 63
Resonance*
ETHAFOAM HS 70

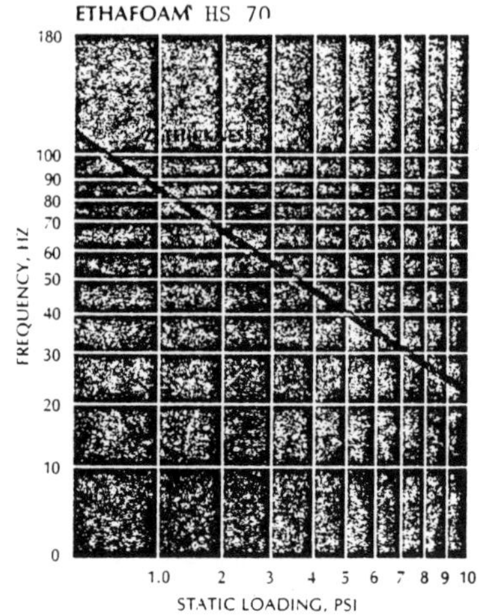

EXPANDED POLYSTYRENE
G. VS. STATIC STRESS – FIRST DROP

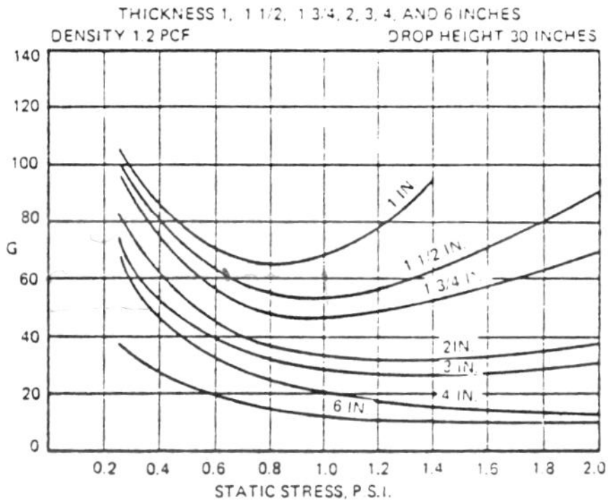

EXPANDED POLYSTYRENE
G. VS. STATIC STRESS – FIFTH DROP

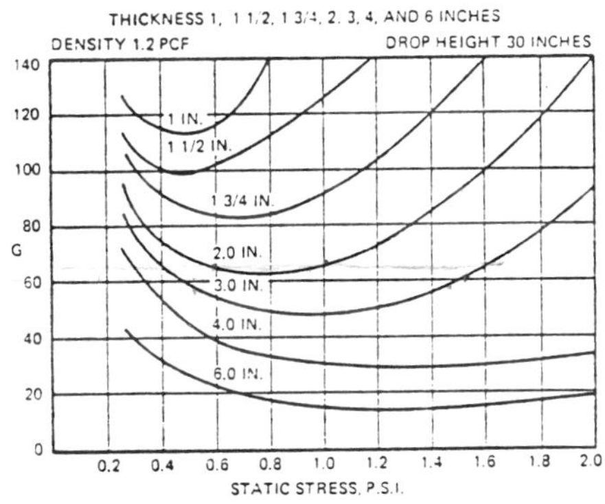

Courtesy of Arco Polymers, Inc.

8. GENERAL CONSIDERATIONS FOR PACKAGE TESTING

8.0 DAMAGE

The main purpose of package testing is to determine whether or not the package can protect the product. If product damage occurs, the package is deemed inadequate; damage is the critical criterion for judging the effectiveness of the package. But what is product damage? Damage can be simple breakage, the shortening of the useful life of a product, or something as subtle as a mar or a scratch. It is important to always precisely define what product damage is prior to the testing of a package.

The treatment that is applied to the specimen for most package testing is supposedly a reflection or a simulation of the physical abuse that occurs in distribution. Our knowledge at present about the nature of these environmental inputs is rather limited, as is the capability of our test equipment to recreate these conditions in the laboratory. However, good testing equipment allows us to create controlled and repeatable inputs in the package testing laboratory. For instance, a good shock machine will produce a variety of shock pulses time after time. This constant treatment may then be used as a reference point to correlate the results of lab tests with the survival of the package in distribution and product damage. With experience the packaging technologist learns to choose the appropriate testing treatment for the packaged product.

PROBLEM SET, CHAPTER 8

1. An article falls from a 10 ft. high stack and hits a piece of linear cushioning (k = 800 lb/in).

   A. How many seconds does it take to hit the cushion on the floor.

   B. At what velocity will it be traveling when it strikes the cushion?

   C. What is the minimum velocity change that the product-package will experience?

   D. What is the maximum?

   E. Can you predict the duration of the shock pulse resulting from this impact? If so, what is it?

   F. What is the maximum shock (in g's) resulting from this impact?

2. A product sits at rest upon a piece of linear, undamped cushioning material. The static deflection is measured to be 0.12 in. The product and its cushion are placed on a vibration table which is set to vibrate at 7.5 Hz with a single amplitude maximum displacement of 0.5 inch.

   A. What is the resulting single amplitude maximum displacement of the product?

   B. What is the maximum acceleration experienced by the product during the vibration?

   C. If the product-cushion system is dropped cushion side down from a height of 36 inches, what shock (in g's) will be experienced by the product?

3. An article weighing 38.5 lbs. can sustain a maximum shock of 45 g's. Find the spring constant for a linear, undamped cushion which will limit the acceleration on the article to 45 g's in a six foot drop. What is the maximum deflection of the cushion? What is the duration of the shock pulse sustained by the article?

   If this article and cushion were dropped from any height, at what frequency would the article vibrate on the cushion after impact (ignore damping)?

4. In testing we find that a 10 lb. product has a critical velocity of 110 in/sec and a critical acceleration of 86 g's. We wish to protect the product for multiple drops from a maximum drop height of 42 inches. We then select a cushion and place the product and cushion on the vibration table. The product on its cushion is found to resonate at 24 Hz. Will the product be adequately protected from shock (at h = 42 in.) by this linear, undamped cushion?

5. Cushion Material A is known to be a linear cushion. We take a sample piece of this material and mount it on the vibration table with a 50 lb. weight on top of the cushion. We start the table vibrating and increase the frequency until resonance occurs at 9.9 Hz.

   A. What is the spring constant, k, of Cushion Material A?

   We take a piece of Cushion Material A and place it on top of a piece of Cushion Material B. Linear Cushion Material B has a spring constant of 333.3 lb/in.

   B. What is the equivalent spring constant of these two cushions acting together?

   We place a 20 lb. product on top of our cushion system as shown below and drop the system 45 in.

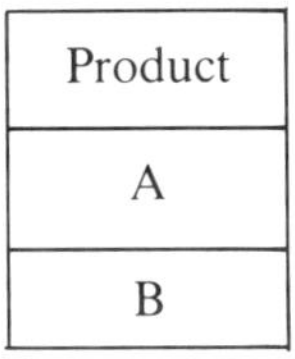

   C. What is the shock, G_m, at the product? What is the shock duration at the product?

   The product is known to contain a critical element with a natural frequency of 37.6 Hz and a G_s of 35 g's.

   D. Was the critical element damaged in the 45 inch drop?

6. A product sits at rest upon a piece of linear, undamped cushioning material. The static deflection is measured to be 0.35 inches. The product and its cushion are placed in a truck which vibrates at 4 Hz with a maximum (zero to peak) acceleration of 0.5 g.

   A. What is the maximum acceleration experienced by the product?

   B. What is the resulting (zero to peak) maximum displacement of the product?

   C. Suppose the product weighs 35 lbs.; what is the shock, in g's, to the product if the product and the cushion are dropped 30 inches onto another piece of the same cushioning material? (Ignore the weight of the cushion and assume enough cushion is present to prevent bottoming out.)

7. In handling a 20 lb. product-cushion system is observed to suffer a 50 g shock from a drop height of 38 in.

This product is particularly sensitive to vibration. It has been found that accelerations exceeding 2 g's at frequencies between 10 and 20 Hz will cause damage. If this product-cushion is transported in a vehicle which has a forcing frequency of 16 Hz at a maximum acceleration of 0.5 g's, will vibration damage be sustained?

8. A small lightweight mechanical accelerometer has a stylus with a natural frequency of 19 Hz. In calibration, the stylus appeared to register 10 g's/division when the shock duration was 8 ms.

   The device is then firmly attached to a 20 lb. product which has two critical elements. The natural frequency of critical element A is 18 Hz. The natural frequency of critical element B is 34 Hz. The product is dropped from a height of three feet onto a linear, undamped cushion for which k = 650 lb/in.

   A. What is the amplitude (in g's) of the shock experienced by critical element A?

   B. What is the amplitude (in g's) of the shock experienced by critical element B?

   C. What reading (in divisions) does the stylus on the accelerometer device show?

9. A spring 3.6 in. long deflects .098 in. when a load of 24.5 lbs. is placed on it. Find the spring constant and the natural frequency of the spring-mass system. If the spring is now cut in half and the two halves placed side by side, find the new natural frequency.

10. The six 2 ms plastic programmers on a shock machine act like a system of linear springs, limiting the duration of the shock to the table to 2 ms. From what height should the table be dropped in order that it receive a shock of 160 g's?

ANSWER KEY

ANSWER KEY

CHAPTER 2

1. t = 1 sec, v = 32.2 ft/sec, x = 16.1 ft.
   t = 5 sec, v = 161.0 ft/sec, x = 402.5 ft.
   t = 10 sec, v = 322.0 ft/sec, x = 1610.0 ft.

2. 24 ft/sec, x = 8.94 ft.
   55 ft/sec, x = 46.97 ft.
   100 ft/sec, x = 155.28 ft.
   200 ft/sec, x = 621.12 ft.

3. x = 6", t = 0.18 sec; x = 18", t = 0.31 sec.
   x = 35", t = 0.43 sec; x = 60", t = 0.56 sec.
   x = 0.75m, t = 0.39 sec; x = 2.25m, t = 0.68 sec.

4. x = 6", v = 68.1 in/sec; x = 18", v = 117.9 in/sec
   x = 36", v = 166.8 in/sec; x = 60", v = 215.3 in/sec
   x = 0.75m, v = 3.8 m/sec; x = 2.25m, v = 6.6 m/sec

5. 0.53 sec.
   204 in/sec.

CHAPTER 3

1. 12 lb/in.

2. 4.6 lb/in.

3. 36.4 lb/in.

4. 5.8 lb/in. 5.95

5. 0.17 in.

6. 2.6 in.

7. 0.63 in.

8. 3.3 lb/in.

9. 16.6 lb/in.

10. 3.13 Hz, 2.98 Hz

11. 5.29 Hz

12. 1.81 Hz

13. 25.56 Hz

14. 0.016 in.

15. 2.33 Hz

16. A. 2898 in/sec^2
    B. 966 in/sec^2
    C. 5796 in/sec^2
    D. 12880 in/sec^2
    E. 25760 in/sec^2

17. 5.1 Hz

18. A. 12.24 lb/in.
    B. 3.82 Hz

19. 78.4 in/sec
    9.35 g

20. A. 50 lb/in.
    B. 7 Hz
    C. 0.44 in.
    D. 4.72 Hz

21. A. 111.1 lb/in.
    B. 37.0 lb/in.
    C. 8.1 Hz

22. 2.5 lb/in^2, 4 lb/in.

23. 30%, 12 lb.

24. 2.28 Hz

26 y(t) = 1.1 sin (15.7 t)

CHAPTER 4

1. 3 g, 1.44 in., Out of Phase

2. 17.3, 1.73 g, In Phase

3. -0.54, 0.05 g, Out of Phase

4. 0.092 in.

5. 3.6 Hz, 0.58 in., 0.99 g, Out of Phase

6. A. 0.034 in.
   B. 2.29
   C. 0.078 in.
   D. 1.15 g
   E. In Phase

7. A. 1.44 in.
   B. 9.4 g

8. A. 1.98 g, In Phase
   B. 0.39 in.

9. A. 3.13 Hz
   B. 379 in/sec^2
   C. 1.55 g
   D. Out of Phase

10. A. 0.015 in.
    B. 18 Hz
    C. 20.8 Hz, 16.1 Hz

CHAPTER 6

1. 1067 lb/in.
   0.9 in.
   16.9 ms

2. 260.4 lb/in.
   9.6 in.
   221.5 ms

3. 50.4 in. x 50.4 in. x 50.4 in.

4. 17.7 in. x 20.7 in. x 23.7 in.

5. 6.3 in. x 9.3 in. x 5.8 in.

6. 54.2 g
   22.8 ms

7. 48.5 g
   22.8 ms

8. 20.9 in. x 13.2 in. x 20.2 in.

9. 10.8 g/div.

   A. 15.83 div.
   B. 8.70 div.
   C. 2.18 div.
   D. 2.27 div.

10. A shock occurred
    A shock occurred, probably on a corner

11. 9.05 div.

12. 2.42 div.

13. 620 lb/in.
    2.3 in.
    33.7 ms

14. A. 27.5 g, 51 ms
    B. $G_m (2h) = \sqrt{2} G_m (h)$
       $\tau (2h) \cong \tau (h)$

15. A. 25 Hz
    B. 1.65
    C. 165 g
    D. 156 g
    E. 11.2 ms
    F. 1.74
    G. The cushion does not adequately protect the product.

16. A. 94.2 g
    B. 106.2 g

17. 1050 lb/in.
    1.31 in.
    10 in. x 10 in. x 12 in.

18. 32 g

19. 4.41 div.

20. 23.7 g

21. Yes, the article is undamaged.

22. A. 260.4 lb/in.
    B. Yes, adequate protection
    C. 4.38 in.
    D. 13.8 in. x 16.8 in. x 14.8 in
    E. 62 ms

23. A. Not adequately protected
    B. Yes
    C. Adequate protection
    D. No

24. 1875

25. A. 5 g/div.
    B. 180 g
    C. Expect no damage
    D. Yes

26. A. 10 div.
    B. 157 g

27. No

28. 41.6 g

29. Cushion not adequate but no bottoming out

30. 7

CHAPTER 7

1. 252 in/sec., 587 in/sec.

2.

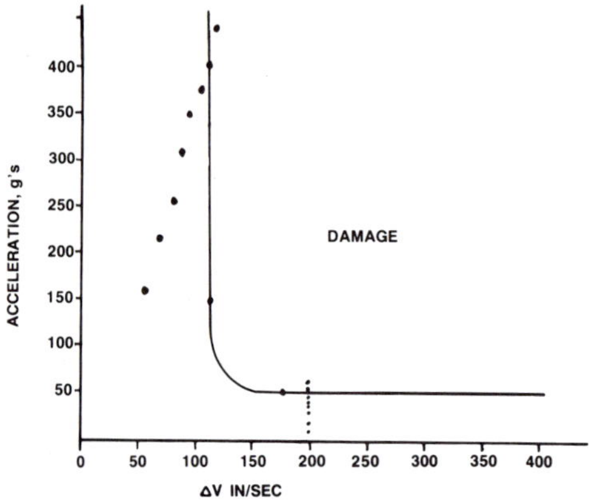

3.

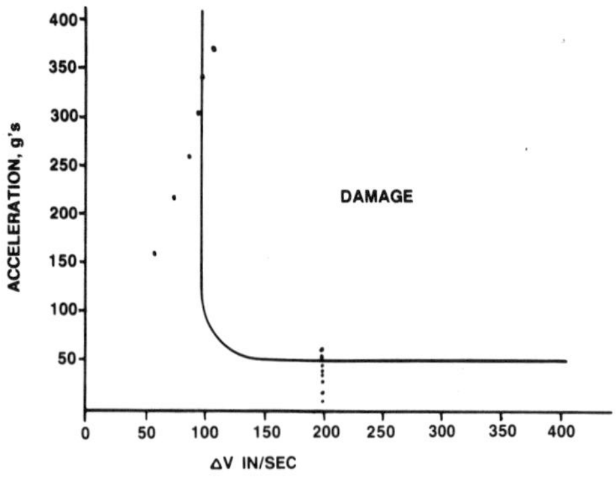

4. 35.7 in^2 of 2 inch thick ETHAFOAM™ 220 (@ 1.4 psi) (use Figure 8 p.

5. 50.0 in^2 of 2 inch thick DYLITE Polystyrene (@ 1.0 psi); or 31.3 in^2 of thick DYLITE Polystyrene (@ 1.6 psi) (use data on p. 133)

6. 87.0 in^2 of 5 inch thick ETHAFOAM™ 220 (@ 2.3 psi) (9" x 9.7"

7. $\Delta V_{max} \cong$ 193 in/sec < V_c = 200 in/sec, no cushioning needed for protection.

8. 31.8 in^2 of 3 inch thick ETHAFOAM™ 220 (@ 1.1 psi)

9. k ≅ 984 lb/in

$d_m \cong$ 1.6 in.

$\tau \cong$ 30 ms

10. We could use 7.1 in^2 of 2 inch thick ETHAFOAM™ 200, but this re approximately 80 Hz, producing a M = 4.3 for the cushion; resulting of a vibration induced acceleration. We could use 7.1 in^2 of 2 ETHAFOAM™ HS70 which resonates at approximately 93 Hz vibration somewhat below the 0.3 g level.

11. Cushion: 6.2" x 6.2" x 3" Polystyrene (38.5 in^2 of 3 inch fo

Container: 15" x 14" x 15"

Weight: 50.5 lb.

12. 595 lb/in

2.06 in.

9.28" x 11.28" x 17.28"

13. 10" x 10" x 1

4.12" x 4.12"

4" x 5"

14. Adequate s

CHAPTER 8

1. A. 0.79 sec
   B. 304.5 in/sec
   C. 304.5 in/sec
   D. 609 in/sec
   E. No
   F. Cannot calculate G_m

2. A. 1.61 in.
   B. 9.25 g
   C. 24.5 g

3. 541 lb/in
   3.2 in
   42.6 ms
   11.7 Hz

4. Yes

5. A. 500 lb/in
   B. 200 lb/in
   C. 30 g, 50.5 ms
   D. Damage expected

6. A. 1.17 g
   B. 0.71 in.
   C. 9.3 g

7. Expect vibration damage

8. A. 75.9 g
   B. 84.6 g
   C. 13.3 divisions

9. 250 lb/in
   10 Hz
   14 Hz

10. 2 in.

3.

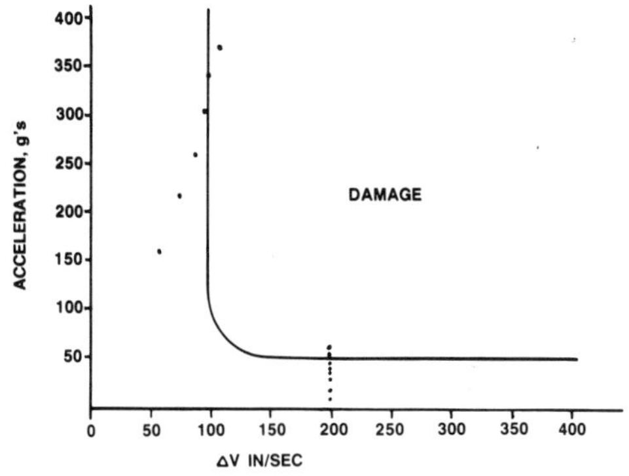

4. 35.7 in^2 of 2 inch thick ETHAFOAM™ 220 (@ 1.4 psi) (use Figure 8 p. 126)

5. 50.0 in^2 of 2 inch thick DYLITE Polystyrene (@ 1.0 psi); or 31.3 in^2 of 3 inch thick DYLITE Polystyrene (@ 1.6 psi) (use data on p. 133)

6. 87.0 in^2 of 5 inch thick ETHAFOAM™ 220 (@ 2.3 psi) (9" x 9.7" x 5")

7. $\Delta V_{max} \cong 193$ in/sec $< V_c = 200$ in/sec, no cushioning needed for shock protection.

8. 31.8 in^2 of 3 inch thick ETHAFOAM™ 220 (@ 1.1 psi)

9. $k \cong 984$ lb/in

   $d_m \cong 1.6$ in.

   $\tau \cong 30$ ms

10. We could use 7.1 in^2 of 2 inch thick ETHAFOAM™ 200, but this resonates at approximately 80 Hz, producing a M = 4.3 for the cushion; resulting in too high of a vibration induced acceleration. We could use 7.1 in^2 of 2 inch thick ETHAFOAM™ HS70 which resonates at approximately 93 Hz, reducing vibration somewhat below the 0.3 g level.

11. Cushion: 6.2" x 6.2" x 3" Polystyrene (38.5 in^2 of 3 inch foam)

    Container: 15" x 14" x 15"

    Weight: 50.5 lb.

12. 595 lb/in

    2.06 in.

    9.28" x 11.28" x 17.28"

13. 10" x 10" x 12"

    4.12" x 4.12" x 2"

    4" x 5"

14. Adequate spring length

APPENDIX A
PIEZOELECTRIC ACCELEROMETERS

APPENDIX A

PIEZOELECTRIC ACCELEROMETERS[1]

An accelerometer is a device that measures acceleration in mechanical shock or vibration. The principle element of a modern accelerometer is a piezoelectric transducer that produces electric charges as a response to mechanical stress.

The internal components of an accelerometer, shown in Figure A1, are very much like a simple spring mass system. The spring is the piezoelectric material and a relatively heavy mass rests on top of it. The mass is preloaded with a mechanical spring to keep it in place and the entire assembly is sealed in a metal housing. When the accelerometer is subjected to an external force, the mass exerts a stress on the piezoelectric material and a charge is produced. As graphed in Figure A2, the sensitivity of a piezoelectric accelerometer is defined as the ratio of its electric output to the acceleration of the external force. Depending on the electronic design of the accelerometer, the output may be an electric charge or voltage:

Output as Voltage (millivolts)

sensitivity = mV/g

Output as Charge (pico-Coulombs)

sensitivity = pC/g

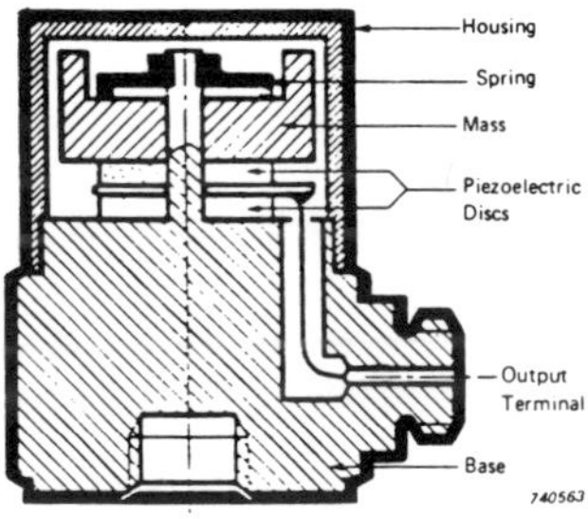

Schematic of piezoelectric accelerometer (Reference 18).

Figure A1

The direct output of a piezoelectric accelerometer is of very low power content and cannot be connected directly to a measuring device for display purposes. A preamplifier conditions and amplifies the output signal of an accelerometer to a level that is suitable for the measurement or display devices. There are voltage preamplifiers and charge preamplifiers for the two types of accelerometers. The

[1]See Reference 18 for a complete discussion of piezoelectric accelerometers.

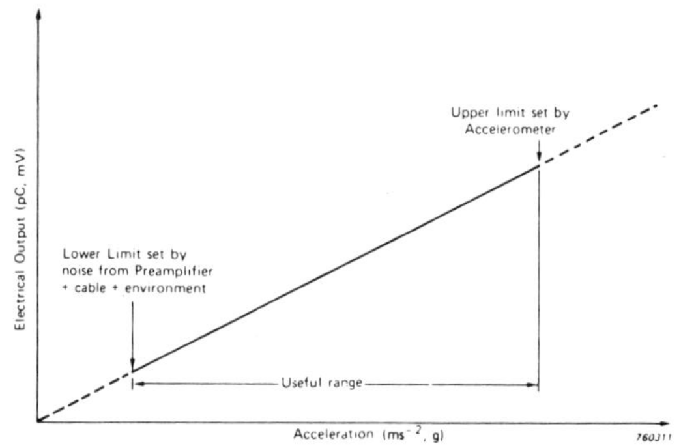

Electrical output versus acceleration for piezoelectric accelerometer (Reference 18).

Figure A2

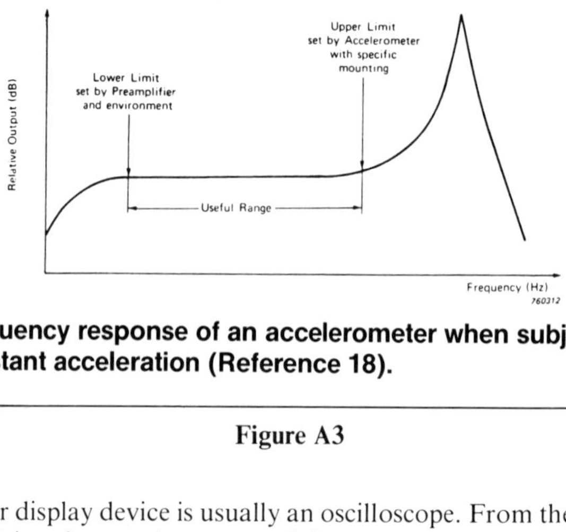

Frequency response of an accelerometer when subjected to constant acceleration (Reference 18).

Figure A3

measurement or display device is usually an oscilloscope. From the user's point of view, the sensitivity of an accelerometer of either type should be described as mV/g for the accelerometer coupled with its own preamplifer.

The output of the accelerometer is not constant over all frequencies of vibration of the same amplitude. Generally speaking, the response is lower at the low frequencies and higher as the external vibration force approaches the natural frequency of the accelerometer, as shown in Figure A3.

The ideal accelerometer should have high sensitivity, maximum frequency range and minimum weight. Unfortunately, it is still in the process of being designed. The following is a list of important parameters to be considered for selecting the right accelerometer for a particular application:

1. Enough sensitivity to measure the smallest force you have in mind. The maximum force should not exceed 1/3 of the maximum rating of the accelerometer.

2. The output of the accelerometer-preamplifier is compatible with your measuring device.

3. Estimate the frequency range required for your measurement and make sure that this lies within the linear frequency range of the accelerometer.

4. The mass of the accelerometer should be at least ten times less than the mass of the specimen.

5. Make sure that the accelerometer functions properly under the test environment (high temperatures, rapid temperature changes, etc.).

CALIBRATION

The calibration of an accelerometer is basically obtaining its sensitivity under different parameters such as frequency response, temperature response and amplitude linearity. New accelerometers are calibrated at the factory and the calibration chart is included as part of the specification of the accelerometer. With use, or misuse, recalibration is often necessary. Sensitivity and frequency response should be recalibrated at 6 or 12 month intervals. It is also recommended that recalibration be performed prior to the start of extensive tests. There are two categories of calibration methods:

Comparison Method

The unknown accelerometer is mounted on top of a reference standard accelerometer (or as close as possible) and both of them are subjected to controlled vibration. The sensitivity of the unknown accelerometer is obtained by comparing the output of both accelerometers, as shown in Figure A4.

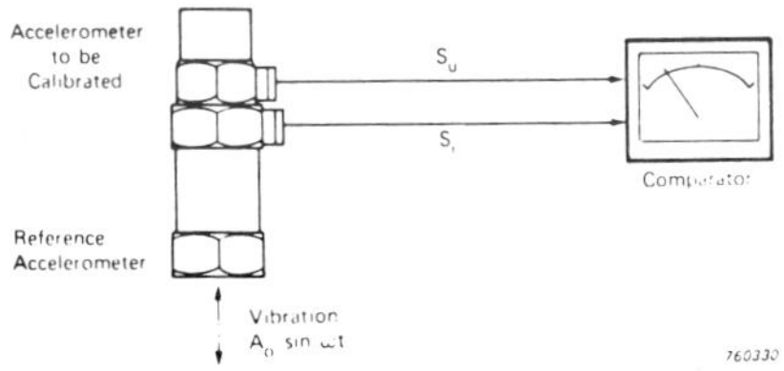

Back to back comparison calibration of accelerometer (Reference 18).

Figure A4

Absolute Method

When an object is undergoing sinusoidal vibration, the relationship between the maximum acceleration and the maximum displacement is:

$$a_{max} = A\,(2\pi f)^2$$

where f is the frequency of the vibration. For calibration, we subject an accelerometer to a sinusoidal vibration with known frequency and take accurate measurement of the maximum displacement during vibration. The acceleration experienced by the accelerometer can be calculated by the above relationship of the sinusoidal motion, and the measurement of frequency and displacement. With modern lab equipment, an accuracy of ± 0.5% is obtainable.

APPENDIX B
ASTM TEST STANDARDS

APPENDIX B
ASTM TEST STANDARDS

D 642-76 (1983)	Compression Test for Shipping Containers
D 775-80	Drop Test for Loaded Boxes
D 782-82	Shipping Containers in Revolving Hexagonal Drum
D 880-79	Incline Impact Test for Shipping Containers
D 959-80	Drop Test for Filled Bags
D 997-80	Drop Test for Loaded Boxes
D 999-75 (1981)	Vibration Testing of Shipping Containers
D 1185-73 (1981)	Pallets, Testing
D 1596-78a (1983)	Shock Absorbing Characteristics of Package Cushioning Materials
D 2221-68 (1984)	Creep Properties of Package Cushioning Materials
D 2956-71 (1977)	Controlled Shock Input Tests for Shipping Containers
D 3331-77	Assessment of Mechanical-Shock Fragility Using Package Cushioning Materials
D 3332-77 (1983)	Mechanical-Shock Fragility of Products, Using Shock machines
D 3580-80	Vibration (Vertical Sinusoidal Motion) Test of Products
D 4003-81	Controlled Horizontal Impact Test for Shipping Containers
D 4169-84	Performance Testing of Shipping Containers and Systems

REFERENCES

REFERENCES

1. Butkunas, A.A. 1967. "Power Spectral Density and Ride Evaluation." Sound and Vibration, December 1967.

2. Goff, J.W. and Twede, D. 1983. Shake and Break, Laboratory Adventures in Package Dynamics. Published by Goff and Twede.

3. Green, Kent C. 1983. "New Measurement and Analysis Techniques for Packaged Product Testing." IBM Technical Report TR 02.1072, October 12, 1983.

4. Guins, S. G. 1981. Notes on Package Design. 4th Edition.

5. Halliday, David and Resnick, Robert. 1966. Physics Parts I and II. John Wiley and Sons, Inc., New York.

6. Harris, Cyril M. and Crede, Charles E. 1976. Shock and Vibration Handbook, Second Edition. McGraw Hill.

7. Hewitt, Paul G. 1974. Conceptual Physics, A New Introduction to Your Environment. Little, Brown and Company, Boston, MA.

8. Kerr, Mark T. 1981. The Importance of Package Testing in Today's Data Processing Industry. Proceedings of Western Regional Forum Packaging Institute, USA.

9. Kerr, Mark T. 1982. "Random Vibration for Package Testing." IBM Technical Report TR 02.995, May 1982.

10. Mindlin, R.D. Dynamics of Package Cushioning. Bell Telephone System. Technical Publications, Monograph B-1369.

11. Newton, R.E. 1976. Fragility Assessment,Theory and Test Procedure. L.A.B. Report 160.06.

12. Ostrem, Fred E. and Godshall, W.D. 1979. An Assessment of the Common Carrier Shipping Environment. General Technical Report FPL22. Forest Products Laboratory, Forest Service, U.S. Department of Agriculture, Madison, WI.

13. Pennington, Dale. Basic Shock and Vibration Theory. Endevo Corporation, Dynamic Instrument Division, TP 219. Reprinted from Environmental Quarterly.

14. Pierce, S. R. and Goff, J. 1970. Control of Damage in Shipment. Project 1, Multi-Sponsor Research Program, School of Packaging, Michigan State University. August 1970.

15. Tevelow, Frank L. The Military Logistical Transportation Vibration Environment: Its Characterization and Relevance to MIL-STD Fuze Vibration Testing. HDL-SR-83-11. The Journal of the DoD Fuze Engineering Standardization Working Group, U.S. Army Electronics Research and Development Command, Harry Diamond Laboratories.

16. Thomson, William. Theory of Vibration with Applications. Second Edition. Prentice-Hall, Englewood Cliffs, New Jersey.

17. Winne, R. J. 1977. Packaging Report F-7741. "What Really Happens to Your Package in Trucks and Trailers." The Packaging Institute, U.S.A.

18. 1976. Piezoelectric Accelerometers and Vibration Preamplifiers Theory and Application Handbook. Bruel and Kjaer Company. May, 1976.

19. 1978. Military Standardization Handbook. Package Cushioning Design. MIL-HDBK-304B. U.S. Department of Defense.

20. 1981. The Fundamentals of Signal Analysis. Hewlett Packard Application Note 243, November, 1981.

21. 1984. Annual Book of ASTM Standards. American Society of Testing and Materials.

22. 5 Step Packaging Development. L.A.B.

INDEX

INDEX